Through Technology, It is Now Possible for Anyone to Contact the Other Side

Do we continue to survive after death? Is it possible to talk to those who are supposedly dead? This book provides evidence that death is only a gateway to another level of existence and even more amazingly, this book will show you that the so-called dead can and do communicate with those still on earth. You will learn how to do this on your own and without the help of a medium.

Why would someone want to contact the other side? There are more reasons than there are pages in this book, but proof of survival and knowledge about the afterlife would certainly be a good reason. Another reason people wish to communicate with the other side is simply because they have a loved one who is there. They long to know that their loved one continues and is all right. Perhaps they wish to hear words of love or maybe there were things that were left unsaid. Many people simply do not like to rely on faith or what others say, and only believe in something through direct experience. Whatever the reason for seeking contact, the desire to reach those who are no longer on earth is as old as the human race.

In this book, you will read the stories of many individuals who have received messages from those in other dimensions through common electronic devices. Perhaps reading about these conversations from other worlds will make you want to experience these phenomena for yourself. If so, this book will show you how. You will learn that it is possible for anyone to contact the other side by using a simple audio recorder, video recorder, or computer.

Dedication

This book is dedicated to the members of the American Association of Electronic Voice Phenomena (AA-EVP). If they had not shared their stories with the original founder of the AA-EVP, Sarah Estep, and then later with us, this book would never have come to be. This book is a celebration of the achievements of AA-EVP members in the ever growing field of Electronic Voice Phenomena and Instrumental Transcommunication. The book is also dedicated to all of the pioneers in this field who have generously shared their research with the public.

Acknowledgments

We wish to say a special thank you to Barbara Thurman, Sarah Estep and her daughter Becky, Janice Oberding and Tracy Sherwood for their assistance and support in making this book a reality.

Warning

Although the publishers have made every effort to insure the information was correct at the time of going to press, the authors and publishers do not assume and hereby disclaim any liability to any party for any loss, damage or injury caused by information contained in this book. Further, there are potential hazards to the mentally delicate associated with EVP/ITC experiments, as the results may not be what the person expects. The authors, publisher and agents disclaim any liability resulting from use of this book.

There is No Death

And

There are No Dead

**Evidence of Survival and Spirit Communication
Through Voices and Images from those on the Other Side**

By

Tom and Lisa Butler

AA-EVP Publishing

PO Box 13111, Reno, Nevada 89507

First Edition

Fourth Printing, 2008

Website: www.aaevp.com

This book is available from Internet book sellers, by order from local book stores, from the authors at www.book.aaevp.com and by writing:

AA-EVP
PO Box 13111
Reno, NV 59507

ISBN 0-9727493-0-6

Cover design by Tom and Lisa Butler.

Printed in the United States on acid-free paper by Lightning Source, LaVergne, Tennessee, www.lightningsource.com

Table of Contents

About the Authors

Lisa has a Bachelor's Degree in Psychology and Tom has a Bachelor of Science Degree as an Electronics Engineer. They assumed the role of Directors for the American Association of Electronic Voice Phenomena (AA-EVP) in 2000 after the founder, pioneer EVP researcher, Sarah Estep, retired. They publish a quarterly NewsJournal for Association members and have published articles in other magazines and journals. They are also the Directors of the Department of Phenomenal Evidence for the National Spiritualist Association of Churches.

Both became interested in survival, phenomena and metaphysical concepts at an early age. They have been working in the field of Electronic Voice Phenomena and recording the voices for over fourteen years. Their current research is focused on receiving paranormal features through Video Instrumental Transcommunication. When not writing or conducting research and experiments, the two conduct lectures about nonphysical phenomena for various groups.

To Write to the Authors: If you wish to contact the authors or would like more information about this book or the American Association of Electronic Voice Phenomena, you can send a self-addressed stamped envelope or an international postal reply coupon to:

AA-EVP Publishing
PO Box 13111
Reno, Nevada 89507
USA

Or send email to membership@aaevp.com

The authors appreciate hearing from you, learning about how you feel about this book and if it has helped you. The authors would also enjoy hearing about your experiences with contact from the other side. Your experiences may be shared in the AA-EVP NewsJournal, articles or in another book. Due to the amount of correspondence, AA-EVP Publishing cannot guarantee that every letter written to the authors can be answered.

About This Book

This book provides an extensive history of Electronic Voice Phenomena (EVP) and Instrumental Transcommunication (ITC). You will come to understand that these phenomena have been documented over a longer period of time than you might have first realized. The first part of this book contains chapters with stories and examples of EVP and ITC. These are just a few of many such stories. The number of people that have experienced these communications adds to the validity that these paranormal voices and pictures are real occurrences. Most of the people who collect these phenomena will tell you that they were not born with any special gifts, and that anyone can receive the voices and pictures with a little patience and persistence. In the second part of the book, you will find detailed information on how to conduct experiments to record the voices and images from the other side.

The Book Cover

The cover design of this book includes a feature we collected during a Video ITC experiment in which we asked our contacts on the other side to give us an image for the book cover. They did give us one. From EVP messages we have received, we believe the face on the cover is one of our guides. The image has been enhanced and the background color changed for the printing of this cover; however, the original image can be seen at

http://aaevp.com/resources/pictures_in_no_dead.htm

Clarification of Terms

Artifact: An artifact is something that is produced by electronic equipment or the environment. An artifact may be such common things as sound in an audio recording caused by electrical static or a light flair in film or digital photography. A mundane artifact may be mistaken for nonphysical phenomena; however, sound and light energy caused by these natural disturbances may be used by the communicating entity to create messages or images.

Entity: Any self-aware being is an entity. Since we do not always know if the entity communicating with us is a discarnate person, we often use the more generic word, "Entity," to describe the "person" we are communicating with from the other side.

Nonphysical: There is just one reality, but reality has many aspects. The Physical Plane is an aspect of reality. Nonphysical is used in this book to denote the aspects of reality that are not considered part of the Physical Plane. Some of the other terms used to denote the nonphysical aspect of reality are, "The other side," "Other dimension" and "Beyond the veil."

Classification of EVP. To provide a tool with which people could specify the quality of their EVP samples, Sarah Estep defined a grading system using the following three levels.

Class A: Voices can be heard and understood over a speaker by most people. *(The average person will hear what you expect them to hear <u>without</u> prompting.)*

Class B: Voices can be heard over a speaker, but not everyone will agree as to what is said. *(The average person will hear what you expect them to hear <u>with</u> prompting.)*

Class C: Voices must usually be heard with headphones and are difficult to understand. *(There is usually no sense in playing Class C EVP for someone else.)* Class C or B voices may have one or two clearly understood words. However, loud does not equal Class A.

A Note about References and Quotes

An attempt has been made to provide ample information about our sources so that you can, at the very least, see where we have begun our research. In these days of the Internet, many important references are published only on the Internet. Web pages are often changed, so we understand that this is not a firm reference, but the reference should at least be a starting point for retracing our steps. When referring to websites, we have omitted the http:// from the beginning of the web address except where necessary for clarity. This will automatically be inserted by your web browser when you enter the web address.

Reference numbers are shown as a superscript number, and if a particular page of a book is being cited, that will be shown in parentheses. For instance, David Wilson.[45(118-123)] This is for Reference Number 45:

1. Rogo, D Scott and Bayless, Raymond—*Phone Calls From The Dead*, 1979 Prentice-Hall, Inc., New Jersey.

Most of the stories and examples came from articles in the AA-EVP NewsJournal. Because of the complexity of trying to reference the NewsJournal issue and page that each of these stories appeared in, we have decided to make a blanket statement that, if the source is not given, you should first look to the AA-EVP NewsJournal. Sarah Estep published seventy-four NewsJournals before we assumed leadership of the Association and it is through her work and documentation that you are able to see many of the stories in this book. In some special instances; however, we have referenced the AA-EVP NewsJournal and other Journals, in doing so we have shown the particular Volume and Number as: [54(V2N4)]

The comments that come from those in other dimensions are printed in *italics*. By their nature, EVP and ITC do not lend themselves well to demonstration through the written word. The paranormal pictures do not print well, if at all, and with EVP, you will want to hear the sound of the EVP voices. At aaevp.com, you can find EVP examples that you can listen to with your computer and full-color ITC images that are better able to be seen than the printed version. You will find many other examples on the Internet. If you do not have access to the Internet, either on your computer or through a friend or the library, the next best solution would be for you to record EVP for yourself. We are confident that you would soon have plenty of examples of your own.

Part I

Electronic Voice Phenomena
And
Instrumental Transcommunication

Our role model for our experimentation in Electronic Voice Phenomena (EVP) and Instrumental Transcommunication (ITC) has been the experimenters and researchers who have come before us. It is they, from whom we have learned the methods by which we may experience these phenomena, what has been learned by their trial and error, and the breadth of what seems possible. In that same light, we offer you here the history, stories and examples of EVP and ITC so that you might also learn, experience and understand these life-changing phenomena.

Techniques and theory on how to record EVP and ITC are provided in Part II.

Chapter 1

A History

A New Way to Communicate With the "Dead"

From the earliest recorded history humans have sought ways to communicate with the "dead." Often, this was through a shaman or an oracle. In more recent times people communicated with those who had passed to the Spirit World through mediums. With the rise in popularity of mediums like John Edward, we now see this type of spirit communication broadcast to millions of television viewers.

What is not as well known is that, in the last hundred years, communication with the so-called dead has also been taking place through various mechanical and electrical devices. Today people record the voices of the unseen on electronic and digital devices and some experiments have even been carried out using lasers.

This possibility of spirit entities recording their voice on tape is beyond the comprehension of most people. But, as you will see from this chapter on the history of such communication, the so-called dead have used our advancing technology to communicate at increasingly sophisticated and evidential levels.

At first there was no real term for these paranormal communications. Then in the 1970s, the voices were given the name: Electronic Voice Phenomena, or EVP. The term, EVP, signified the recording of the voices on audio tape. As technology further progressed, the communications from those unseen sources in other dimensions became even more fantastic and were not just recorded voices. Various researchers received pictures, computer messages, faxes, and even two way communications via radio and telephone. The new term, Instrumental Transcommunication, or ITC, came into being to address this expanded spirit communication.

When a loved one crosses to the other side, most of us wish that we could once more communicate with them. Perhaps there are things that we did not take the time to tell them or there are things that we wish we could ask. The need to communicate often comes simply from the desire to know that they are all right. It might be wrong to assume this need is only felt on our part. The discarnate might also have the need to contact their loved ones who now reside in a different place from where they are. It certainly seems that this is the case.

The history of spirit communication using electronic devices is filled with information indicating that many of the technical designs for creating these devices have come through mediums or the mediumship of the inventor. It seems that we are not the only ones experimenting with instruments to make contact across the veil. Throughout history, the discarnate also seem to be experimenting in techniques to contact us.

One of the reasons the history is presented here is to show that these paranormal communications have not only occurred over a long period of time, but they have occurred in many different ways to many different people living around the world. People who have substantial training in the sciences have experienced these phenomena. With the depth and complexity found in the way these phenomena occur, we believe you can see reason to accept that EVP and ITC are not isolated phenomena experienced by just a few fortunate people. EVP and ITC represent real communication that is frequently demonstrated. At the same time, these factors also indicate that communication can be received by nearly anyone who is willing to make a committed attempt.

A History of EVP and ITC

Waldemar Bogoras. Bogoras was exiled to Siberia as a youth where he had the opportunity to observe the natives of eastern Siberia. He also lived in the United States for several years of his life. Bogoras received praise from his peers for his artistic gifts, scientific insight, power of observation, descriptive clarity and careful analysis of observed facts. He also wrote novels under the pen name of Tan.

In approximately **1901**, Bogoras observed a Siberian shaman perform a conjuring ritual.[17(99)] In a darkened room, the shaman beat the drum more and more rapidly while entering into a trance state. Bogoras was astonished by the strange voices that he heard filling the room. They seemed to come from everywhere and were speaking in both English and Russian.

Bogoras set up his recording equipment so that he could record in the dark with the shaman sitting twenty feet from him. The light was turned off, and after some hesitation, the spirit voices were once again heard. They followed the shaman's request that they speak directly into the horn of Bogoras' portable Edison phonograph.

The Bogoras recording is the first in which the "direct voices" of spirits were registered on a recording device. The recording showed a clear difference between the spirit voices, which seemed to come directly from the mouth of the phonograph horn, and the voice of the shaman some distance away. Throughout the recording, "...the shaman's ceaseless drum beats can be heard as if to prove that he remained in the same spot."

=====0=====

Dr. J. L. Matla and **Dr. G. J. Zaalbert van Zelst.** Documented attempts were made in Holland in the early **1900s** to prove that spirit entities could manipulate devices. Dr. J. L. Matla and Dr. G. J. Zaalbert van Zelst began research into this area in 1904.[45(138-139)] Both men were physicists and over a long period of time had investigated psychics and mediums in their hometown. Most of their experimental equipment was designed from information that had come through the mediums that they had been studying.

Their first attempt included a small cylinder that was hermetically sealed. Air could escape from the cylinder through a rubber tube connected to a device they called a "Manometer." The manometer looked like a carpenter's level and had an alcohol bubble that floated in a glass tube. The two scientists also built a special room in which the experiments took place. It was sealed to provide a controlled environment in which drafts, tremors and other physical influences could not cause readings on the manometer. Tests in the room could be monitored through a window in an adjoining control room.

The two wanted to see if they could encourage a spirit to influence the air in the cylinder, thus causing a movement of the alcohol bubble. During their experiments, the men sat in the control room mentally invoking the unseen entities. They reported that a deflection was almost immediately registered by the manometer in their first experiment. They conducted further tests using two cylinders and were successful in their requests for the discarnate to manipulate one cylinder while not affecting the other. In their book, *La Mystere de la Mort* (*The Mystery of Death*), they stated that there was no normal explanation for the consistent results they received. Both were convinced that unseen entities were the cause of the phenomena.

They next built a more complex device called a "Dynamistograph." An indicator, a key and a register were its only working parts. The indicator consisted of a wheel that was electrically driven and on which letters of the alphabet were written. When activated, the wheel rotated sequentially. The wheel was connected to a very sensitive key, and when the key was depressed the letter at the top of the wheel was printed on the register. An old-fashioned Wimshurst Machine powered the device through static charges. Meaningful messages were recorded from the device for over a year, even when it was left alone and the two physicists were not there.

The men discovered that their results suffered when it was raining or humid and their best communications were received during the drier periods of the year. Also, certain electrical currents channeled through the device produced better results and Matla came to the conclusion that, "The very element of our personality that survives death is partially electrical in nature and has an affinity for manipulating electrical energy."

=====0=====

David Wilson. In **1915,** David Wilson,[45(118-123)] a London amateur wireless operator, designed and built a detection device that was sensitive to electrical influences. It was attached to a galvanometer, which registered the presence and strength of an electric current. Wilson saw movement of the galvanometer over several days and finally noticed that the meter seemed to be making organized movements. He wondered if this could be Morse Code. A few days later, on June 10, 1915, the galvanometer registered activity for eight continuous minutes. Wilson was able to translate the registrations into Morse Code. The message read, *"Great difficulty; await message, five days, six evenings."*

Wilson reported his findings in the March 13, 1915 issue of *Light.*[46] He decided that he needed an independent observer to verify what he was receiving. When the day came for the experiment with the observer, Wilson was unsure if the machine would even work, as the registrations of the device had become incoherent. He was relieved and astounded when the dial recorded Morse Code for almost half an hour. The letters were taken down independently by both men. When the two were compared it was apparent that the message read, *"Try eliminate vibrations. ARTK"*

Wilson modified his device based on the understanding of this message and added a human subject into the design along with a dummy Morse key. In Wilson's words, "The only possible way in which the message could come through the receiver would be by means of an agency which could not only affect the new detector but also the brain of the ... 'circuit person.' Moreover, these actions would have to be synchronous before the needle of the galvanometer would deflect." Wilson received some communication with the new device, but the contents were disappointing. He reconstructed his machine several more times finally removing the deflecting needle and linking the device to another apparatus that converted all incoming code into audible signals that represented dots and dashes.

Wilson collaborated with a friend who lived in Paris to see if the same device, separated by a large distance from the first, would pick up the same message. He gave the friend instructions and a duplicate device was created. On March 19, 1916, Wilson's machine received the message *"Nyet leezdyes Kogoeedbood kto gavoreet poroosky?"* In English, *"Is there anyone who speaks Russian here?"* Six minutes later, a somewhat distorted and fragmented message was picked up by the machine in Paris. It recorded, *"Nyet ... lee ... (incoherent).... Kto ... porooski."* After publishing these findings in *Light,* Wilson dropped from sight never publishing anything on his work again.

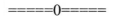

Grace Boylan. In 1918, Grace Boylan's book, *Thy Son Liveth: Messages from a Soldier to His Mother,* was published by Little, Brown and Company.[12] Grace and her son, Bob, had both learned Morse Code. Bob would often practice using Morse Code by sending messages to his mother who was in another part of the house.

During World War I, Bob, a second lieutenant, was sent to France. Grace was reading a letter from him when the wireless indicated that a message was coming in. The message was from Bob and read, *"Mother...I am alive and loving you, but my body is with thousands of other mothers' boys near Lens. Get this fact to others if you can. It's awful when you grieve, and we can't get in touch with you to tell you we are all right. This is a clumsy way. I'll figure out something easier. I'm confused yet. Bob."*

A later message came from Bob telling his mother of his death and asking her to tell others that there was no "horror in death." He told her that he had been in the middle of battle when suddenly another Lieutenant had touched his arm and said, "Our command has crossed, let's go." At first Bob did not understand, but then realized that they were all dead. He told his mother that he had then sent the Morse Code message from an enemy's wireless station. All later communication from Bob took place through automatic writing.

=====0=====

Thomas Edison. Thomas Edison believed that there could be a radio frequency between the long and short waves, which would make possible some form of telepathic contact with the other world.[44] In the October **1920** issue of *Scientific American,*[47] Edison was quoted as saying, "If our personality survives, then it is strictly logical or scientific to assume that it retains memory, intellect, other faculties and knowledge that we acquire on this earth. Therefore, if personality exists after what we call death, it is reasonable to conclude that those who leave the earth would like to communicate with those they have left here. I am inclined to believe that our personality hereafter will be able to affect matter. If this reasoning be correct, then, if we can evolve an instrument so delicate as to be affected by our personality as it survives in the next life, such an instrument, when made available, ought to record something."

There has been much speculation and rumor that Edison did work on some sort of telephone that would reach the "dead," but no plans of such a machine have ever been found. A blueprint of a purported Edison device did surface in New York in the early 1940s; however, the blueprint was not felt to be authentic. A machine was constructed from the blueprints, but it did not work.

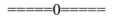

F.R. Melton. F.R. Melton was an inventor and psychic investigator in England.[45(125)] His son, George, had been a wireless operator during World War I. George had received anomalous messages over the wireless during that time that could not be accounted for. He had wondered if these messages were coming from discarnate entities. After the war, the Meltons decided to explore this possibility in more detail and were excited in 1920 when they read reports in British newspapers about wireless stations across the country picking up strange and unaccountable signals.

In **1921**, an article written by Melton was published in *Light*.[46] In it, he claimed that he had invented a "psychic telephone." The machine was a telephone connected to an amplifier that was placed in a small box. Melton claimed that he had received many paranormal voices directly over the device. He also published a booklet, *A Psychic Telephone*, which provided illustrations and details on its construction.

Melton did not feel that everyone could use his psychic telephone. He only claimed that the device would amplify psychic voices. His son was a medium and the first device was built specifically to amplify the independent voices that he received.

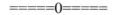

Francis Grierson. Francis Grierson, a United States author and medium, invented a telephonic device through which he received several communications from the deceased. The communications were described in a short book, *Psychophone Messages,* published in Los Angeles in **1921**.

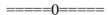

Oscar D'argonell. In **1925,** Brazilian Oscar D'argonell, described contacts with the deceased through the telephone in a book titled, *Voices from Beyond by Telephone.*

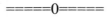

Edgar Wallace. Author Edgar Wallace, used a 78 r.p.m. record cutter for his work. After his death in **1932**, his secretary used the record cutter to cut a disc. When she replayed the record she was shocked to hear a few words spoken by Edgar Wallace. No one took this incident seriously but it may have actually been one of the first recorded paranormal voices.[45 (137)]

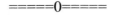

World Broadcasting Company. On April 23, **1933**, a test was set up at the World Broadcasting Company's studios (became Decca Records) in Manhattan, New York, to test the validity of the voices received during a direct voice séance.[54(V7N4)] World studios had the latest sound recording equipment installed. The medium was William Cartheuser and many well-known researchers in the field of parapsychology were in attendance. Among these were Hereward Carrington and Mrs. Helen Bigelow.

The engineers at World were very skeptical and did not believe that voices from the so-called dead could actually be received. They devised test conditions that they felt would eliminate all possibility of fraud. Three microphones were set up for the séance. Microphone 1 was placed on the floor. Microphones 2 and 3 were installed on the opposite corners of the ceiling twenty feet in the air and twenty feet away from the sitters. The microphones on the ceiling were only sensitive to sounds that were within twelve inches and directly in front of them, assuring that they could not pick up voices that would be recorded on Microphone 1. Further, each microphone had a direct connection into the control room.

The medium and sitters were unaware of how the microphones had been set up by the engineers and proceeded with the séance. A spirit voice quickly showed interest in the experiment saying, *"We think they have worked out a very interesting testing procedure for us on their equipment."* The engineers in the control room asked who was speaking. The voice, speaking into microphone 1, told them that he was an engineer in the Spirit World who had colleagues with him. He then told them that they, the spirits in attendance, were all interested in cooperating with making a recording.

The engineers in the control room requested that the person, calling himself a spirit and who was now speaking in Microphone 1, to speak directly into Microphones 2 and 3. Almost immediately, the voice answered within inches of both mics. The voice said that they wanted to provide a demonstration and proceeded to make a quick circuit of the three microphones while speaking a simple short sentence.

Next, the voice introduced a colleague and told those present that he was an eminent research engineer in the science of sound. The spirit research engineer told the sitters and studio engineers that he and the other communicators were, *"Surviving personalities speaking to you from another dimension."* He then moved his voice from the normal level of the human male voice, 300 Hertz, to levels of 3,000 to 5,000 Hertz. While doing this, his voice trailed off to sound like, "An incredibly distant radio signal." The spirit engineer then descended through the frequencies so that eventually his voice sounded like, "a giant mumbling at the bottom of a well." And finally, "Like the lowest note on the longest pipe in a giant organ."

The original spirit voice returned and thanked the sitters and engineers for helping in the experiment. He concluded with an offer of help and collaboration in future tests. The records of the experiment were sent to the American Society for Psychical Research. However, the recording stayed at World Broadcasting, as they did not want to testify publicly that spirit voices had been recorded in their studio.

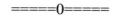

Attila von Szalay. In **1938** Attila von Szalay, a natural psychic, heard a voice calling his name. The voice yelled, *"Art"* and von Szalay was certain that he recognized it as that of his deceased son. Von Szalay continued to receive these "direct voices" and in 1941 attempted to record them on a 78 r.p.m. record cutter with little success. In **1947** von Szalay began working with a magnetic tape recorder and received some voices of clear quality.

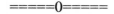

Harry Gardner and **J. Gilbert Wright.** In the **1940s**, Harry Gardner and J. Gilbert Wright[45(129)] invented a device that was similar to the Melton telephone. Gardner claimed that the apparatus could act as a channel for spirit voices in the presence of a psychic. The device was simple; a twenty-four-inch by seven-inch box was lined with sound-proofing material. A microphone was placed inside the box and a small hole was drilled in the side of the box that led to a loudspeaker. The box was later lined with finely reduced iron to make it magnetic. In the initial experiments, voices immediately issued through the loudspeaker even with the device several feet away from the medium.

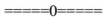

N. Zwann. The *Psychic News,*[7] a British Spiritualist publication, reported that N. Zwann had come to England in **1947** with plans for a spirit radio. The plans had been received through mediumistic communication. *Psychic News* also reported that the use of the apparatus had been successful. In 1949 the Spirit Electronic Communication Society was formed in Manchester, England, and Zwaan demonstrated a device named "Super Rays." It was later renamed "Zwaan Rays" in honor of Zwann.

=====0=====

Leslie Flint. In **1948** during a demonstration at Denison House, London, an American wire recording apparatus was used to produce gramophone records of the spirit voices made possible by the direct voice medium, Leslie Flint.[84] The recording was a success, and the direct voices were recorded as clearly as when they were spoken by the various spirit entities.

=====0=====

Father Ernetti and Father Gemelli. In Italy in **1952**, Father Ernetti was collaborating on music research with Father Dr. Gemelli[68(12)] in the Experimental Physics Laboratory of the Catholic University of Milan. Ernetti was a respected medical doctor and nuclear physicist and Gemelli was the President of the Papal Academy. The two were conducting oscillographic experiments with a wire magnetophone (wire recorder) in an effort to produce clearer singing voices in Gregorian chants. The wire broke frequently and it was the habit of Father Gemelli to call on his paternal father when things went wrong in his day-to-day activities. That day when the wire broke once again, he called out, "Oh father, help me!"

After the experiment, the two men played back what had registered on the magnetophone and to their astonishment, heard a voice say, *"But of course I'll help you! I'm always with you."* Gemelli recognized the voice as that of his father. They turned the magnetophone back on and Germelli asked out loud, "Papa, if you are really here, please repeat what you said before." On playback the reply followed immediately after his request with *"But Zucchini* (a childhood nickname only known by his father), *it is clear, don't you know it is I?"* These messages were later reported to Pope Pius XII.

Over thirty years after this event, in October of 1986 Father Ernetti broke a long silence and publicly reported in the Italian magazine, *Oggi* on his experiments with a group of physicists in the construction of a "Chronovisor," a device that permitted reception of pictures and sounds of events from the past.[54(V5N4)] Ernetti reported that they had been successful with the device dating back to the 1950s.

Father Ernetti's original theory was based on accepting one of the principles of classical science, which predicated that light and sound waves are not lost after emission but transformed and remain present.[69] He felt that it theoretically would be possible to reconstitute them by restoring them to their original energy pattern.

The Chronovisor consisted of three elements. First a series of antennas were linked in a chain. They were formed out of different alloys and received all or nearly all of the electromagnetic and non-electromagnetic waves existing in space. The second element was designed to convert what the antennas received into an electronic signal. The third element produced sounds and images similar to a television set. The device was said to have picked up early pictures and sounds from the ancient days of Rome and from other places and times.

=====0=====

Raymond Bayless and Attila von Szalay. Raymond Bayless became interested in Attila von Szalay's earlier work with spirit voices in **1956** and the two men decided to resume von Szalay's earlier experiments.[45(85-90)] They wanted to determine if von Szalay really could produce objective psychic voices and if they could physically isolate and separate von Szalay from the source of the voices. The surprising outcome of the experiments led to the first discovery of the tape-recorded voice phenomenon. They published their initial findings as a letter in the January 1959 issue of *The Journal of the American Society for Psychical Research.*[48] It was a few months after this that Friedrich Jürgenson[13] announced the discovery of the voices in Sweden.

The two constructed a cabinet that von Szalay sat in while trying to generate the voices. A microphone was placed in the opening of a trumpet, a device used by Spiritualist mediums to amplify spirit voices, and then placed in the cabinet. A tape recorder was placed outside the cabinet and connected to a speaker so that any voices or noises developed within the enclosure could be heard. Whistles, whispered voices and rapping were heard coming from the speakers whether von Szalay was in the cabinet or outside of it several feet away. The voices could be recognized as male and female and often sounded mechanical.

Bayless reported to the ASPR journal that the first taped voices happened on December 5, 1956. Attila had been sitting in the cabinet, Bayless was outside of the cabinet observing and both men felt that nothing had taken place. When the recording was reviewed neither man expected to hear anything; however, they were astonished when they heard a voice clearly say, *"This is G."* They immediately performed another test. They stood outside the cabinet and in view of each other. Attila made single whistles while Bayless listened to test the amplifying system. Bayless suddenly realized that they were receiving low whistles in reply. The December test demonstrated that von Szalay was not only able to produce these paranormal voices but was also able to get them to appear directly on the tape recorder.

More tests were performed, and on the first pilot test, three distinctive voices were heard but were so garbled they could not be understood. It may be of interest to readers that Bayless wrote, "I've encountered this type of reception with von Szalay very often. The psychic voices will be almost as loud as our own, but so 'mush-mouthed' that they cannot be understood." Most researchers remain baffled to this day as to why some taped voices are often difficult to understand even when they are quite loud.

A very evidential voice, that gave credence to the supposition that the voices were coming from discarnates, was recorded on July 7, 1957. Von Szalay was experimenting alone. He had heard no independent voices but on listening back to the recording of the session he heard, *"Hot dog, Art!"* The voice ended with a high pitched laugh that had a special meaning for von Szalay. He recognized the voice as that of a woman that he had dated during the Depression. The two had lived on two-for-a-nickel hot dogs during these hard times and had promised never to forget their hot dog dinners.

The researchers soon realized that the voices showed intelligence. One experiment gave the location of Bayless' brother after being asked where he was. The brother had left town without telling him where he was moving, the information was later verified as correct. When asked the name of Bayless' grandmother, a name unknown to von Szalay, a voice answered, *"Emma,"* which was correct. Once, von Szalay recognized the recorded voice of a woman he had known years earlier. They later learned that she had committed suicide only days before her voice was recorded.

At first the recorded voices were only five to seven words, but over many years of research, the length of the messages increased and one message of forty-five seconds was recorded. The voices gave their names, delivered messages, answered questions and sometimes even gave advice on better methods of recording and equipment setup. Bayless and von Szalay commented on the criticism, that the voices were merely random radio broadcasts, as ridiculous by pointing to the above evidence. They noted that the entities often used crass profanity, something that would not be picked up on the radio. They also felt that the single most important thing that they had been able to conclude from their research was the fact that the voices were somehow independent of Attila von Szalay's mind.

======0======

Friedrich Jürgenson. Friedrich Jürgenson,[13] of Sweden, recorded his first voices on a tape recorder in **1959**. He was recording the sounds of bird songs near his villa. He was startled to hear among the bird's melodies a voice talking about, "nocturnal bird songs." He wondered if he had picked up a stray radio broadcast but felt that it was quite a coincidence to have picked up those words, as he was recording birds singing. He did more recordings and heard his deceased mother's voice saying, *"Friedrich, you are being watched. Friedel, my little Friedel, can you hear me?"* He devoted himself to the recording of the voices and published several books, *Voice Transmissions with the Deceased,*[13] *Voices from the Universe,* and *Radio-Link with the Beyond.*

Jürgenson had very close connections with the Vatican and Pope Paul VI. In 1969, the Pope gave Jürgenson the Commander's Cross of the Order of St. Gregory the Great. The award is given to acknowledge an individual's meritorious service to the Church. Jürgenson told others that he had found a sympathetic ear for the voice phenomenon in the Vatican.

Jürgenson crossed to the other side on October 15 1987, and appeared on the television set of Claude and Ellen Thorlin just a few days later. Ellen, a very psychic individual, heard an inner voice repeating, *"Channel 4,"* and felt that it had something to do with Jürgenson's funeral.[54(V7N2)] Claude got out his Polaroid camera and shortly before the 1 p.m. funeral, turned the television on. Channel 4 was an empty station in Sweden and so it simply displayed random noise. They watched this snow for quite a while and had almost given up. Ellen even left the room, but then a strange thing happened. A spot of light appeared and expanded quickly, disappeared and then expanded again. Claude released the camera shutter. The Polaroid picture developed into the face of their friend Jürgenson. The picture was taken at the same time of Jürgenson's burial service being held seventy-five miles away.

=====0=====

Stewart Robb. Another interesting communication with the other side was reported as taking place in England in the **1960s**. Stewart Robb tells about this experiment in his book *Strange Prophecies that Came True.*[60] Robb writes that Michael Ash had told him about a laboratory in Surrey, England where he witnessed a new way to communicate with the next world. Robb asked Ash to write an account of what he saw and he wrote the following, "The setup was a radiation source in a lead screen. The radiation from this source was being recorded by a Geiger counter onto a cathode ray oscilloscope as a spiral tracing. The tracing consisted of a record of each pair of charges produced by the disintegration of the radiation source. The pattern was photographed and was found to be made up of short and long signals like international Morse Code. The message was decoded by observing that the signals at the start of a letter were of a slightly different duration than those at the end of a letter. Anyone who knew international Morse Code could thus decode the signals recorded, and records were kept of the words thus produced.

"There was evidence of considerable education (although the person communicating could not spell very well). The information received was highly technical and had to do with improvements in the setup used and instruction to the user ... The messages were signed "M.F.," making Michael Faraday, a famous physicist who died in the 1900s, suspect as their possible source. The messages came in at the rate of near eight thousand words a minute, as each photographed spiral contained great numbers of dot and dashes."

=====0=====

Konstantin Raudive. Dr. Konstantin Raudive, a psychologist, philosopher and Latvian then living in Sweden, read one of the Friedrich Jürgenson books. Raudive was intrigued with the voice phenomena, but skeptical, and asked if he could join Jürgenson for a recording session. After working with Jürgenson, he became convinced that the voices were real. Jürgenson taught Raudive how to record and from **1965** on he devoted his time to the voice recordings. Both Raudive and Jürgenson were multilingual and the voices they recorded were a mixture of languages. These voices were unlike any normal voice broadcast. The speech was almost double the usual speed and the sound was pulsed in rhythms like poetry or chanting.[49(226)]

Many engineers, scientists and experts worked with Raudive over the years conducting voice experiments. Physicist, Professor Alexander Schneider, was one of them. In 1969 Raudive and Professor Schneider were jointly given the first prize awarded by the Swiss Association for Parapsychology for their work on direct voice messages on tape recordings.

Although not the first person to record EVP, Raudive is given a good deal of credit for being the first to bring Electronic Voice Phenomena to the attention of a larger audience. His book, *The Inaudible Made Audible,* was translated into English in **1971** and published by Colin Smythe, Ltd. under the title *Breakthrough: An Amazing Experiment in Electronic Communication with the Dead.*[2] In the preface to *Breakthrough,* Smythe writes that before publishing the book, he wanted to be sure that the voice phenomenon was real. He did some test recordings and thought that he heard a voice, but he could not understand it. He asked Peter Bander, the editor of *Breakthrough,* to listen to the tape. After listening, Bander heard a woman's voice say in German, *"Why don't you open the door?"* Bander recognized it as his mother's voice. Bander and his mother had done all of their correspondence by tape and her voice was unmistakable. The message also made sense, because during the previous week, Bander had insisted on keeping the door of his office closed and his colleagues had teased him for his seclusion. Bander knew that Smythe could not understand German and so asked others to write down phonetically what they heard. They all heard the same thing.

The voices became known as "Raudive Voices" after *Breakthrough* was published. However, Colin Smythe and Peter Bander became more aware of Friedrich Jürgenson's role and continued activity in voice phenomena research. It was obvious to them that a less personal and more accurate name needed to be coined for the phenomena. Peter Bander used the term, "Electronic Voice Phenomena," in the introduction to his book, *Carry on Talking.* Smythe said that their policy to use the term, "Electronic Voice Phenomenon," in an official sense was first carried out in a determined fashion in an article written by Malcolm Hughes in *The Spiritualist Gazette,* in April of 1973.

In **1971**, controlled EVP experiments were conducted with Raudive by the chief engineers of Pye Records, Ltd.[50(59-63)] Precautions were taken to prevent freak pick-ups of any kind. Controls within the experiment also excluded random high or low frequencies being received. Raudive was not allowed to touch the equipment and was allowed only to speak into a microphone. No one present heard anything but Raudive speaking while the recording was being made. However, when the recording was played back, over two hundred voices were found on the eighteen minutes of tape. Many of these messages were personal and very evidential to those who were there. In his book, *Carry on Talking,* published in 1972, Peter Bander said that there was so much excitement from those who were there that the experiments continued into the early hours of the morning. *Carry on Talking* was published in the United States as *Voices From the Tapes: Recordings from the Other World.*[50]

In **1972,** Belling and Lee, Ltd., at Enfield, England, conducted experiments with Raudive and the recording of the paranormal voices in their Radio Frequency Screened Laboratory.[50(65-67)] Peter Hale supervised the experiments. Peter, a physicist and electronics engineer, was considered the leading expert on electronic-suppression in Great Britain. The Belling and Lee lab was used to test the most sophisticated electronic equipment for British defense and was expressly designed to screen out electromagnetic transmissions. Before the experiment, Hale had expressed his opinion that Raudive's voices originated from normal radio signals. The lab's own recording equipment was used for the test and paranormal voices, that should not have been there, were recorded on factory fresh tape. Peter Hale said after the experiment, "I cannot explain what happened in normal physical terms."

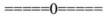

Franz Seidl. In **1967,** Franz Seidl of Vienna, developed a device called the "psychophone."[45(142)] This consisted of a primitive type of radio receiver with a wide frequency range combined with an amplifier. A tape recorder was attached to this and any voices coming over the receiver were automatically recorded. Siedl described his device and his work in, *The Phenomenon of Transcendental Voices,* published in 1971.

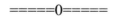

Marcello Bacci. Marcello Bacci, of Grosseto, Italy, has been experimenting in the paranormal for many years and is still active today. In the early **1970s** he developed direct "Electro Acoustic Voices" that conveyed long messages via a tube radio and were capable of partial dialogue. The people who were lucky enough to attend a Bacci session often heard their departed loved ones talk directly to them through Bacci's radios.[31] Also in the **1970s,** associations of people interested in EVP were formed in Germany, Italy, and Austria.

=====0=====

George Meek, Bill O'Neil and **Dr. Müller.** Americans George Meek, Hans Heckman and Paul Jones opened a small laboratory to conduct research into the phenomena of EVP in **1971.** Meek, a retired engineer had been interested in survival after death for many years. The idea for building a device to speak to the so-called dead was given to him by a discarnate scientist during a séance. The discarnate scientist told Meek that he would cooperate in building the device by giving the laboratory team instructions. Meek wanted to achieve two-way communication with the other side and became convinced that sophisticated equipment needed to be developed if communication with those on the other side was to improve.

In 1977, Meek was introduced to Bill O'Neil, a gifted medium and electronics engineer. O'Neil's spirit communicator called himself "Doc Nick" and told him that he was a former radio ham operator. Doc Nick told O'Neil that the development team should try using certain audio frequencies instead of the white noise that was being used by most researchers. Doc Nick delivered technical information on how to build the communication device and told the researchers that it would provide thousands of sensitive frequencies that the other side would be able to use for communication. A series of devices were made.

Soon, another spirit was in contact with O'Neil. His name was Dr. George Jeffries Müller. He materialized in O'Neil's living room and said that he had come to join the team in their work on the device. In October of 1977, Dr. Müller's first words were recorded on the device now called "Spiricom." Müller gave the team considerable personal information about himself. He told them that he died in 1967 and had been a college professor. He gave them his social security number and advised them where to find his death certificate. All of the information was verified as correct.

Meek held a press conference on April 6, 1982, at the National Press Club in Washington, D.C. He told those who attended, "An elementary start has been made toward the eventual perfection of an electromagnetic-etheric communications system, which will someday permit those living on earth to have telephone-like conversations with persons very much alive in higher levels of consciousness."

Tapes of the conversations between O'Neil and Müller were made available to the public at the press conference. They provided very interesting listening, ranging from mundane discussions on food to technical advice on how to build experimental video equipment. The sound of Müller's voice is fascinating as it sounds much like a robot in a science fiction movie. The "robotic" sound of Müller's voice was an artifact of the suite of audio frequencies that were used in Spiricom.

Meek had hoped to present the device to the media at the press conference, and by doing so, reach a large public audience. However, a large section of the media refused to attend. In the end the conference made little impact and Spiricom went largely unreported. In addition, Müller eventually told Bill O'Neil that he needed to move on and the two-way conversations via Spiricom ceased.

One of the limitations of Spiricom was that it seemed to be dependent on William O'Neil. George Meek had others try out the device, one being American researcher and founder of the AA-EVP, Sarah Estep. The fantastic results achieved by O'Neil were never repeated.

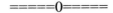

Scott Rogo and **Raymond Bayless**. In **1979**, two American parapsychologists, Scott Rogo and Raymond Bayless, published their book, *Phone Calls from the Dead.*[45] The book was the result of a two-year investigation into phantom phone calls. It provided empirical evidence that many people had received telephone calls from loved ones after the person had transitioned to the other side, and that such phantom phone calls were more common than the authors originally believed. These calls from another dimension were of short duration and did not register on the telephone company's equipment. There were even reports of calls coming through telephones that were not connected.

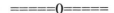

Manfred Boden. In **1980**, German cabinetmaker, Manfred Boden, was at his computer when letters and entire lines of text on his computer monitor began to change on their own. Boden's first name and family name appeared, then the name of an acquaintance that had died three months before. The message read, *"I am here. You will die, Manfred, 1982 accident, August 16, 1982. Yours, Klaus."* A later message changed and listed the cause of death to be from a heart attack. The message caused Boden much stress, as he was overweight and not living what one would call a healthy lifestyle.

Boden's telephone was also affected. In 1981, strange cracking noises were heard during his telephone calls. Then his telephone conversations became filled with voices of unknown origin. The voices often could only be heard by one of the parties and this led to many misunderstandings. This continued to happen over a period of four years and telephone company traces found no evidence of problems with the phone line or crosstalk.

The various computer contacts and paranormal phone calls received by Boden were investigated by Professor Ernst Senkowski, Dr. Ralf Determeyer, Dr. Theo Locher and Guenter Heim. These men reassured Boden that Klaus, the deceased friend, would never scare him in such a way. The message had to come from a deceptive spirit and impostor. Boden called friends to assure them that he was still alive after the date given for his death had passed. The experience had caused considerable duress.

Unfortunately, researchers can be contacted by those on the other side who are not particularly advanced. This message, however unfriendly, was received on a Commodore CBM 8032 personal computer, and is considered the first known instance of a spirit using a computer to contact a living person.

Manfred Boden's contacts continued and he was able to reach spirits who were more advanced. Researcher, Ernst Senkowski, played an eighteen minute two-way conversation between Boden and several entities who said that they were pure energy at an AA-EVP conference in 1985.

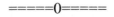

=====0=====

Hans-Otto Koenig. In **1982** electronics engineer, Hans-Otto Koenig, had been closely following George Meek's work. Koenig was an electronics and acoustics expert and created a device using extremely low beat frequency oscillators with ultraviolet and infrared lights. The "Koenig's Generator," as it was called, was set up for a live broadcast on Radio Luxembourg. The station engineers closely supervised the set up of the equipment and Koenig was not allowed to touch the device. The equipment was connected to a set of speakers and switched on. An engineer asked Koenig if the voices would come on request. To everyone's total astonishment a clear voice replied, *"Otto Koenig makes wireless with the dead,"* and chaos ensued. The voice replied, after a second question was asked, saying, *"We hear your voice."*

Rainer Holbe, the program presenter, assured the audience while they were still on the air, "I tell you, dear listeners of Radio Luxembourg, and I swear by the life of my children, that nothing has been manipulated. There are no tricks. It is a voice and we do not know from where it comes."[51(339)] The engineers at the station later issued a statement that they had found no natural explanation for the voices that were heard in the studio and on the air during the live radio broadcast.

Fidelio Koberle reported on Koenig's Radio Luxemburg session in the *VTF Post*[53] newsletter (German EVP association called the Vereins Fur Tonbandstimmenforschung or VTF). He wrote, "Now there are microphone recorded voices of unexpected strength, precision, clear and noise-free. People can no longer say that one is hearing something in the background noise that isn't there."[54(V2N4)]

Hans-Otto Koenig demonstrated his equipment at a conference put on by the VTF in 1984. One person who was present commented, "One can hear the answers from the other side at once so this really is like a call with the telephone. Two mothers that had lost their children called on them and the children answered with clear voices and several times with long sentences."[54]

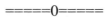

Sarah Estep. In May of **1982**, Sarah Estep founded the American Association of Electronic Voice Phenomena (AA-EVP) after proving to herself that the paranormal voices were real. She published a quarterly NewsJournal to Association members that provided ideas on equipment and experiments, kept them up to date on the latest developments in EVP and ITC, and reported on what various experimenters were receiving in the way of messages. Sarah wrote an outstanding book on EVP and her personal experiences in working with the phenomena titled, *"Voices of Eternity."*[3] It was published in 1989.

=====0=====

Kenneth Webster. In 1984 Ken Webster, and his friend Debby Oakes, reconditioned a house built on a very old foundation in Dodleston, near Chester, England. Not long after they did this, they began experiencing poltergeist activity. Most of it was focused in the kitchen area. The activity included moving furniture, stacked dishes and even hand written notes.

Webster, an economics professor, brought home a computer to work on. By modern standards, the computer was a primitive BBC Model 4. The computer had around 32K of memory with a word processor installed on a chip. The only way to save information was to place it on a floppy disc on an external drive. There was no modem, no Internet and no computer network.

The couple accidentally left the computer on one evening, and when they returned home, they found a poem that neither of them had written on the screen. They felt that this was some kind of joke and assumed that someone must have entered the house while they were away. The next time a message was received, they were certain that the house was secured. A friend offhandedly suggested that they reply to the message. They did and the results were astonishing.

One computer message said "… You live in my house, with lights, which the devil makes. It was a large crime my house to have stolen." Over an extended period of communication, the couple learned that the messages were coming from a man named Thomas Harden. Harden gave information on the period of time he lived in and answered questions about that time period correctly. The language the messages were written in has been called correct for the 16th century by some and incorrect by others.

They asked how Harden communicated with them and were told that it was through a box of lights which sat near his chimney. In another message Harden wrote, "You say you are from 1985. I thought you were from 2109 like your friend who brought me the box of lights."

When Harden, in his time period, was arrested, Webster and Oakes decided to try to contact the beings from 2109. They typed out a message that read, "Calling 2109, calling 2109." Communication was established through the computer to these "energy beings" from the future who had succeeded in synchronizing the time of Webster and Oakes in 1984 with the time of Harden in 1546.

Webster documented the experience in his book *The Vertical Plane.*[52] All together there were more than two hundred and fifty two-way computer contacts over sixteen months.

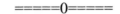

Klaus Schreiber. Klaus Schreiber invented a device in **1985** called, "Vidicom," after studying George Meek's Spiricom design. He had suffered the deaths of many loved ones in his life. After hearing about transcommunication on the radio, he and some friends conducted an experiment as a joke. Nothing was heard until the very end, when an extra voice was picked up. After that, Schreiber continued recording and picked up the voices of his daughter, Karin, and other deceased family members. The EVP voices mentioned using video as part of his experimentation. He did and eventually recorded faces on his television by recording blank channels with a video camera. The technique used by Schreiber consisted of aiming a video camera at the television set. The output of the camera was then feed back into the television thereby creating a feedback loop. The loop created a swirling cloud-like appearance in which the spirit images gradually appeared over several frames. This procedure for receiving video ITC pictures is still used by many researchers and is further described in Chapter 13.

Schreiber's first images were blurred, but over the ensuing years and by using the feedback loop, his pictures greatly improved. He captured a picture of his daughter, Karin, and the faces of other relatives. His daughter became his research counterpart on the other side and assisted him with receiving further contacts from deceased family and friends. Schreiber also received some pictures of those who were famous such as the actress, Romi Schneider. Parapsychologist, Professor Hans Bender, declared that the phenomenon experienced by Schreiber was real.

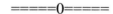

Maggy Harsch-Fischbach and Jules Harsch. Maggy Harsch-Fischbach and her husband Jules Harsch, of Luxembourg, began to receive spectacular voice contacts through radio systems early in their experiments in **1986**.[17(14)] Often a high-pitched, computer-like voice came through their radios to announce the beginning and end of experiments. The entity gave the couple information on how to customize their equipment for better communication. The voice identified itself as a higher being with no name and told them that they could call him "Technician." They soon came to know that there was a group of beings working on the other side to make the contacts possible. They called themselves the "Timestream Research Group."

The Director of Timestream was a scientist named Swejen Salter.[17(10-14)] She told the couple that she was a scientist who had lived in a parallel world and had never lived a life on earth. Salter made frequent telephone contacts to the Harsch-Fischbachs. As time passed many eminent ITC researchers from earth joined the group after their death, including Konstantin Raudive, Friedrich Jürgenson and Klaus Schreiber.

On October 4, 1986, the couple received their first video image from Timestream. It was recorded from a television screen with a video recorder and a video camera. The image was of Pierre K. They had previously received audio transmissions from Pierre. Many more images from the Spirit World came through via video. By 1993, the television transmissions were replaced by computer contacts. Timestream was able to access the hard drive of the couple's computer, leaving detailed messages and amazingly clear computer-scanned images.

The contacts received by the Harsch-Fishbachs were nothing short of phenomenal. The information received covered many topics about how those who were communicating with the couple lived in this other realm of existence. The communication included pictures of people and scenes from this other reality being downloaded into the couple's computer. The main message of all communication was that we do not die and that we continue to live beyond physical death. Information was often of a spiritual nature and also gave information how people should live their life while on earth. Information was also received on the importance of ITC researchers coming together to create a resonant field of energy, and of being of like mind in order for the contacts of these higher beings to continue and improve.

=====0=====

INIT. The International Network for Instrumental Transcommunication (INIT) was formed out of this desire to form a resonant field between researchers here and those on the other side. INIT had its first meeting in September of **1995**. Fifteen ITC researchers attended the meeting that was closed to the public and the media. They formed INIT and signed a declaration of their underlying aim to ensure that ITC spread with an ethical/moral base as well as a technical one.

Information and news of the Harsch-Fischbach contacts were covered extensively in the *Cercle d'Etudes sur la Transcommunication*[62] or *CETL* journal that was translated into English by Hans Heckman. Later more information on the Harsch couple's results, as well as information on what those in INIT were receiving, was published in *CETL INFOnews,* translated by Heckman and edited and published by Mark Macy. Mark continued to provide information on many outstanding and sensational ITC contacts from various researchers in his later journals *Contact* and *Transdimsension*. The book *Conversations Beyond The Light.*[17] written by Dr. Pat Kubis and Mark Macy further spread the word of these amazing contacts.

=====0=====

Consideration of the Evidence of Timestream. As it goes for individuals or groups who receive extraordinary phenomena, the Harsch couple's results came under attack and criticism. This has occurred with Spiricom, Kenneth Webster, The Scole experiments and others. It seems that it only takes one person to write or say something about a researcher's results possibly being fraudulent or questionable, for the word to spread. So we researched the subject via telephone calls and letters to people who had been involved and who are respected as being critical and unbiased observers.

In August of **1988**, respected Swiss parapsychologist, Dr. Theo Locher, was able to experience one of the Harsch couple's contacts with the other side. On the day that Dr. Locher was to visit, Maggy Harsch-Fischbach received a telephone contact from their spirit team that said that contact would be attempted at 8 p.m. when Dr. Locher would be present.

Dr. Locher had an opportunity to look over the experiment room before the attempted contact. He noted that the small black and white television set used for communications was plugged in but that the screen would not light up. The speakers on the set were also not working. At 8:22 p.m., the television screen came to life and he, along with the Harschs, saw a picture of a group of trees for almost three minutes. He also heard the chopped and synthesized voice of Technician clearly from the television set loudspeaker, speaking about the development of ITC contacts. Dr. Raudive greeted them and ended the contact. Dr. Locher was convinced that the phenomenon was genuine. He gave two reasons; the broken television set was made to work and the Technician and Raudive called him by name.

If one goes through just a little of the material on the Harsch-Fischbach research, instances of cross-correspondence are found that seem impossible to have been faked. In one instance, a telephone call from Jürgenson was made, first to Maggy Harsch-Fischbach, and then to Ernst Senkowski. Jürgenson told them that the group on the other side would send a picture through the television of Adolf Homes. That picture did come through Adolf's television and a transcript and picture were received regarding the communication in Luxembourg.

In October of 2002 we contacted eminent researcher and author, Dr. Ernst Senkowski, and asked him about the results that came out of Luxembourg. Quoting from parts of his letter, he wrote that, "There is no doubt from my side that the Harsch-Fischbachs played an important role in the evolution (of ITC) ... I personally witnessed part of these phenomena ... There is no question that the Harschs are psychic and mediumistic. No doubt they realized genuine phenomena which could not have been faked by any normal means. In the case of the Harsch-Fischbachs, I have never seen evidence or proof of fraud.

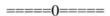

Fritz Malkhoff and **Adolf Homes**. Fritz Malkhoff and Adolf Homes began ITC experiments independently in **1987**. Each received voices on tape. Homes of Rivenich, Germany, placed an advertisement in a newspaper trying to find other people with an interest in the phenomena. Malkhoff of Schweich, Germany, responded to the advertisement and the two began conducting recording experiments together.

They learned of the Harsch-Fischbach contacts with the group Timestream and obtained the schematic for the GA-1 system, which was used by the Harsch-Fischbachs, and had it built.[62] In January 1989, they set up the equipment at the Homes residence and tried it out. They were shocked to hear, *"The souls will lift up to us,"* come out of the radio loudspeaker. On April 4, 1989 a voice came through the telephone, the first of many phone calls from the other side that would be received by the two.

The two set up a Commodore C 64 in an attempt for contact via the computer, and on April 4, 1989, they asked their otherworldly friends for news and stored the text. Two days later, an answer was on the computer. In October of 1989 they began receiving pictures on the television.

In October of 1989 Adolf Homes found a short message on his computer. It read, *"1120 TELELIVINGRM."* He interpreted this as an alert for a video contact at 11:20 A.M. At the announced time, he received the first video picture with a spoken comment. In the next fourteen months, three more pictures were received. The third was of Homes' father-in-law who had died one year earlier.

Between the years of 1989 and 1997, Homes received fifty-three telephone calls from the other side, eighty-four radio or telex messages and one hundred ninety-five computer contacts.[65(98/2)] In 1996, Homes arranged with Deutsche Telekom to have his phone calls monitored and traced for two months. Homes received four paranormal calls during the time of monitoring, while the telephone company registered no calls at all. The first call from the beyond came on January 15 at 3:35 p.m. *"This is Mother .... The results of this telephone control will give assurance to your friends. Mother is going to contact you several times on your phone .... The vibrational ties with your equipment make our contacts possible ...."*

Important Instances of Cross-Correspondence

In 1991 a group of ITC researchers led by Adolf Homes receive an image of Dr. George Jeffries Müller on their television screen. Dr. Müller was the person who carried on two-way conversations with William O'Neil over the Spiricom device created by George Meek. The image was photographed and an eight by ten picture was later presented to Meek.

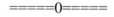

On June 19, 1991, Professor Hans Bender, a German parapsychologist who had made his transition to the other side in May, came through the computer of Maggy and Jules Harsch-Fischbach in Luxembourg and the tape recorder of Adolf Homes in Germany at the same time with the same message.[54(V10N4)]

On June 18, Homes had received a telephone call from the other side with the message, *"Contact tomorrow morning."* The following morning, June 19, he switched on his equipment; two radios, a television set and a tape recorder. He and his wife had to go out on business and left the house. They returned at 10:30 a.m. and found the equipment turned off. The cassette recorder contained a ten-minute recording. The voice was similar to that of Professor Hans Bender's lifetime speaking voice and the message on the Harsch-Fischbachs computer contained the same basic message.

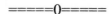

February 10, 1992, Dr. Ernst Senkowski of Germany was working on his computer and referring to a book by Rogo and Bayless, *Phone Calls From The Dead.*[45] The telephone rang and deceased ITC researcher Klaus Schreiber gave his full name. Ernst was on one phone and had his wife pick up an extension phone. Ernst received permission from Klaus to record the call which lasted two minutes. Thirty minutes after Ernst received this call the Harsch-Fischbachs received a call on their answering machine, as they were away from home. The answering machine message referred to the call made to Ernst by Klaus.[54(V11N2)]

On April 28, 1992, a French television crew visited Dr. Ernst Senkowski.[54(V11N3)] The crew conducted an interview and then tried to record EVP voices without convincing results. Right before they were to leave, the phone rang and it was the deceased pioneer EVP researcher, Friedrich Jürgenson, on the line. Permission was granted to record the conversation in which Jürgenson first spoke in French and thanked the television crew. The conversation then continued in German and said that information would also come via the colleagues Homes and Harsch-Fischbach. Ernst wrote, "No outsider knew that the French television crew was here, or planned to be here as the date had been changed shortly before."

The next day Ernst Senkowski spoke with Maggy Harsch-Fischbach on the telephone and she told him that she had had a seven-minute telephone call from Jürgenson before he had placed the call to Senkowski. Jürgenson told Maggy that they planned to send a picture through to the television set of the German researcher, Adolf Homes.

Over a month later, Homes' daughter received a phone call from Raudive, which she recorded.[62(2/92)] Raudive said that a picture of Jürgenson would come through Homes' television the following day. Homes set up his video camera in front of the television the next day, with the television tuned to a blank channel. He turned the camera on and saw a face flash on the screen. At almost the same time a message from Jürgenson was placed on the computer of Maggy and Jules Harsch-Fischbach.[54(V11N3)] The picture received on the Harsch couple's computer showed Swejen Salter, one of the main communicators for the group on the other side called Timestream, as she transmitted the picture of Jürgenson to the television set of Adolf Homes. The picture received on Homes' television was only of the face of Jürgenson and it was the exact same picture that can be seen on the transmitting monitor in the middle of the picture sent to the Harsch-Fischbachs.

On October 13, 1992, Adolf Homes was in his kitchen washing dishes and listening to a musical program on an FM radio station when he heard the words, *"Homes Record!"* over the radio speaker.[54(V11N4)] He grabbed his cassette recorder and microphone and placed them on an armchair in front of the small radio. The music ended and was followed by a news broadcast. Homes remained quiet and suddenly heard, *"This is Doc Müller."* A four-minute conversation then took place between Homes and Müller over the radio. There were two spontaneous contacts between Homes and Doc Müller via the small radio on Oct.13 and Oct.15. On Oct 21 Homes deliberately called on Müller and made contact. All dialogs lasted approximately four minutes. On October 29, Adolf Homes' deceased mother called and told him that William O'Neil was a member of the transcommunication group, Timestream, and was working with Doc. Müller.

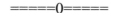

The Scole Group. In October of **1993**, four friends, **Robin and Sandra Foy** and **Alan and Diana Bennett**, sat around a table in a dark cellar, in the small village of Scole, England.[66] Soon, Diana Bennett drifted into trance and a voice spoke through her. The voice introduced himself as Manu and told the group that he was the gatekeeper between earth and the other side. He said that he represented thousands of minds from many other realms of existence. He explained that it was the wish of this group of entities to pioneer methods of communication between the two dimensions using "creative energy" that would be a combination of spiritual, human and earthbound energy forces. He continued, saying that the small group in Norfolk had been chosen to help them. Characteristics of this new form of energy include:

♦ No physical dangers to the health of the medium;

♦ Limitless variety of phenomena produced in this way;

♦ Thought to be relatively quick to develop in a group.

This first contact with Manu heralded experiments that continued over a five-year period. Manu was instructed by a team of scientists and technicians on the other side, and as time passed, the Scole Group achieved incredible breakthroughs that were observed by many senior scientists and investigators. Some of the phenomena that occurred during the sessions were spirit lights that performed intricate maneuvers, responded to verbal requests and were able to go through people and solid objects. Over fifty apports were received by the group. These included various coins and pieces of jewelry. Materialization of solid beings became a common occurrence and these beings were able to be touched by those that were lucky enough to be invited to view the proceedings.

Interesting experiments were undertaken with photography. Handwriting, symbols, pictures and messages were imprinted on factory-sealed, unopened photographic film that had been placed in a locked box. Some of the images were actual photographs of people and places from the past and of other dimensions. Eventually, video cameras were able to record images using blank videotape in a camcorder that was focused on a mirror reflecting a brown ceiling. EVP experiments were conducted via a specially built Germanium device which culminated in an audible performance of Rachmaninoff's Second Piano Concerto, recorded on a machine from which the microphone had been removed. The fact that the Scole Report that was written for the SPR devotes some two hundred and forty pages to the experiments, is an indication of just how varied and extensive the phenomena were.[67]

=====0=====

The history, people and events that have opened the door to EVP and ITC for the world have been described in this chapter. These pioneers have helped us realize the depth and complexity of these phenomena. You will see in the following chapters that people around the world continue the experimentation and research necessary to understand the nature of these phenomena, to improve techniques for communications and to find ways to apply what has been learned for the betterment of humankind.

Chapter 2

Sarah Estep and the AA-EVP

Sarah Estep is the founder of the AA-EVP and continues today to grace us with her wisdom and experience as we continue the work of the Association. We asked her to describe how she discovered EVP and what motivated her to begin the AA-EVP. Her response came to us as an essay titled, *Death No More a Casket*. This title fairly well sums up Sarah's life experiences that brought her to EVP.

Sarah had already come to the conclusion that life ended with death when she was six. Her Grandfather Harry owned a funeral home in another state, and her family visited once a year. In Sarah's words, "Slipping into the viewing room, I softly closed the sliding door behind me. Walking over to the casket, standing on tiptoes, with my hands gripping the edge of the casket, I gazed into the face less than two feet away. Most children would have been terrified but not me. I could tell by looking at the body in front of me, that there was no life there. He couldn't hurt anyone, so I was perfectly safe.

"Two days after the man was buried, a female body was placed in the room for viewing by loved ones and friends. Again, when my family wasn't around, I would slip into the room, close the sliding door and go over to look at her body. My feeling about her was the same as it had been for the man. In my view, there was nothing left in the body. She was gone. She could not hurt anything.

"The day before leaving that summer, I crept into the room again. Leaning as close as I could to the woman who was still in her casket, I wondered if anything came to anyone who died, after death. I knew that my mother and father believed in something they called God, but I decided they'd never seen a dead person as close as I had. There was nothing left in dead people, so obviously they couldn't go on to live another life in what adults called Heaven. The only place a person went at death was into a hole in the ground, where they'd stay as long as our world lasted. Life ended at death's door. Death was a casket.

For a child of six, these were unusual convictions that lasted for over forty years."

Sarah's memory of the grief showed by the loved ones of those silent bodies led her to decide that she would become a social worker after graduation from college. As testament to her ability to focus on a goal, she was soon employed as a social worker in a child welfare agency. She wrote, "During the next four years, I worked with hundreds of children and parents, helping them to a certain extent."

However, it was not until after she married and had three children that she began to have an interest in parapsychology. Her first introduction into the world of the paranormal was through the books about Seth by Jane Roberts.[14] "I started branching out, and turned to books about reincarnation. If we reincarnated then that meant we survived death! How unbelievably wonderful that would be. It would show I'd been totally wrong for most of my life. It also meant that my three deeply loved children would never really die; for them and everyone, death would not be a casket. The books about reincarnation by Dr. Ian Stevenson,[55] psychiatrist, of the University of Virginia, especially appealed to me. I began working in that field since I couldn't accept something of that nature unless I proved it to myself. If I could prove just one child had reincarnated, then I would become a believer.

"A nine-year-old girl I worked with remembered her previous life as an Native American who had died under a tree in the woods while hunting. She also had another amazing factor that strongly suggested she had been her great-grandmother who died several years before she was born. I contacted Dr. Stevenson, and he invited me to bring Mary and her mother to his office in Charlottesville, VA. I did this and we had a good meeting together."

Sarah's work with reincarnation and children came to an abrupt end when the child, Mary, unexpectedly transitioned. However, her exploration into the unexplained aspects of the world continued. As she wrote to us, "Then I read *Handbook of PSI Discoveries*[49] by Sheila Ostrander and Lynn Schroeder. That changed my life.

"The last two chapters in their book focused on recording voices, which were called paranormal, through a tape recorder. They mentioned individuals in Europe, many that were scientists, who had different theories as to how the voices ended up on tape. Some thought they were put there by psychokinesis. Others thought they were a mixture of radio sounds in the environment. A few thought it might be a living person who was able to put his thoughts on the tape. Friedrich Jürgenson,[13] of Sweden, and Konstantin Raudive,[2] of Latvia, were the earliest tapers that had become known for their amazing contacts with those whom they thought once lived on planet earth, and who were now communicating from the Spirit World. These deceased people lived a full life there, and could speak to friends and loved ones through a tape recorder. The authors also mentioned well-known people in this country, such as Walter Uphoff and Harold Sherman,[16] who had listened to some of the European tapes and felt it was a good possibility that the speakers were from spirit."

These books electrified Sarah into action. It is fundamental to her personality that she requires objective evidence before admitting an idea into her worldview as true. So naturally, she had to try to collect EVP samples for herself. "Based on what I'd believed about death for over forty years, obviously something like this—the dead speaking to us from spirit—was impossible."

Sarah committed to experimentation for one week. "The next morning I started, making five-minute recordings as recommended in *Handbook of PSI Discoveries*. I asked the same four questions over and over, allowing time after each question for an answer. Playing it back at the end of the recording, I heard nothing. I taped for two hours every morning, after everyone had left for work or school, and then for an hour at night, when the family went to bed. There was only silence after my questions.

"On the morning of the sixth day, I was so bored with the hours of futile taping that I was ready to quit. Thinking to myself, 'If there is really anyone over there (and of course there wasn't) they must be as bored as I am. Thank goodness I just have twenty-four hours to go before I can stop.' Trying to take away my total feeling of boredom, and theirs (if there was someone), I asked, 'Please tell me what your world is like.'

"In several seconds, a clear Class A voice replied, *'Beauty.'*"

There followed a long period of trial and error as Sarah persisted in her effort to find the best method to record EVP. Other than an occasional encouraging message, such as *"don't give up,"* she recorded mostly whispers and was not at all sure that her initial success was anything more than a fluke. It was 1976 and there were few people experimenting with these phenomena. She communicated with the EVP experimenter, Raymond Cass of England, who suggested that she try using a radio that had air-to-ground frequencies—often referred to as an air-band. Sarah purchased such a radio and was soon recording seven to eight messages a day, many of them what she classifies as, "Class A" messages. This experience also demonstrated for Sarah that a background sound source was needed for the voices to form.

Naturally, Sarah found that using a radio as a background sound source brings criticism from those who would try to discount EVP as an artifact of the equipment, imagination or mistaken interpretation of sounds. About this, she said, "Skeptics of recording voices from spirit are especially delighted when they learn about tapers using sound sources to get voices. They are happiest when they discover many of the best tapers are using a radio as a sound source. 'All of those voices are Earth Plane radio voices from regular broadcasts,' they tell every-one. They can't quite explain why the taper is called by name or is having his direct questions (which can be heard first on tape) answered within a few seconds. Some of us have many contacts on the reverse side of our reel-to-reel and cassette tapes. You'll hear my voice speak-ing backwards, asking a question, or the air-control tower operator or pilot that spoke during the recording, also speaking backwards. You can't interpret what they or I have said. You can only do that if you return the tape to the forward side. Going back to the reverse (wrong) side of the tape, the only thing that is clear is the voice from an invisi-ble. The skeptics will just shake their heads and usually claim there's nothing there, even though—as with the forward side—many are Class A. To me, the reverse voices are some of the best objective evidence we have that we are hearing from another world."

Sarah has had many interesting experiences with EVP. For in-stance, she has had contact with several historical figures and managed to carry on a conversation for over twenty years with an entity claim-ing to be her brother in a previous life. She has also participated with a number of television programs, most of which, concerned field re-cordings in haunted places. These experiences are richly described in her book, *Voices of Eternity.*[3] This book is currently out of print, but there are usually copies available in the Internet used book market.

Interestingly, Sarah reports that she has had very convincing EVP from entities who claimed to originate from other planets in our physical universe. In one instance, they evidently came to visit in person. "One morning they told me, *"We look like yellow. We'll sit, sit by the window."* Thirty-six hours later, at about 8:30 p.m., I was in my office reading, which I do every evening. Looking up, I saw a round yellow sphere the size of a basketball, not ten feet from me. It slowly floated down outside the window. It was in view for about three seconds before vanishing in front on my eyes. I was thrilled. It was too small to be an Unidentified Flying Object (UFO) but I feel it was an object they sent from their ship to me. The next morning during taping, I asked them many questions about it. A loud, clear voice assured me they had sent the object down for me to see. They said, *'We see there. We see after her. Yes, look after her.'* It is obvious that they knew where I sit each night until bedtime.

"Several days later, I developed a serious skin condition. There were large places all over my arms that broke out, first looking like a rash but then developing into unsightly sores. Eventually I went to the doctor who said he'd never seen anything like it. I didn't tell him what might have caused it. He gave me several prescription drugs, and after several weeks it disappeared. I've wondered if I might have had some radiation poisoning."

Small spheres of light are often reported in association with the formation of crop circles. Apparent radiation burns are also occasionally reported by people claiming to have come close to what they believed to be an UFO.

It was not until 1982 that Sarah decided to form an organization that focused on recording voices of the invisibles. She called it the American Association - Electronic Voice Phenomena (AA-EVP) and published her first NewsJournal in May of 1982. Her aim was to help everyone learn to record EVP and to give objective evidence that there was no death.

In the year 2000, and after publishing seventy-four NewsJournals without a miss, Sarah decided it was time to retire from her service to the EVP community. "Eventually, after eighteen years, for several reasons, I knew I would have to give up the AA-EVP. I felt though, that it had to continue in order to help others whether they were taping or not. Who could I ask that would do a good job and who shared basically the same beliefs I had?

"Immediately the names, Lisa and Tom Butler, came to my mind and never left. They had been members for several years, and we had exchanged a few e-mails. We'd never met since they lived thousands of miles from me, and I couldn't claim to know them very well. However, I must have been guided to ask them, for none of the other three hundred names of my membership ever entered my mind. I sent them an e-mail and requested they take over.

"A few days later they called and we talked for a long time. Warm, positive energies became a part of the talk. In an e-mail from Lisa a few days later, she wrote they'd discussed whether to accept leadership and they both felt strongly guided to do this. We worked together and discussed how I would notify members in the May NewsJournal about my leaving and their taking over. That is what I did, and many members joined them. They have done an outstanding job, and I've never regretted for a moment their becoming the new leaders of the AA-EVP. I thank, again and again, the invisibles that guided me into asking them."

To close this chapter, we will quote Sarah concerning her recollection of some of the more important EVP she has recorded. "Whenever I'm being interviewed for a television program, or the radio, I'm always asked what I felt were the most important messages I've ever received. As far as I'm concerned each message is important but there are three that stand out, that are always a part of me. Each was Class A. One of them came as I was finishing my manuscript, *Voices of Eternity*. A clear voice said, *"Your soul is not defeated."* How reassuring that is! It tells us that no matter what difficulties we face as we move through this life our soul will always be the same, it will remain undefeated. We will take with us, as we move to Spirit, the soul we had when we came here. The good, or evil, we have committed during our years here, will determine just what kind of a life we'll find there. We'll have to make up, possibly in many lives to come, what we did to others that may have caused great problems and sadness.

"The second message that is always a part of me came through one morning when I thanked the other side, as I do every so often, for helping me over the years, for communicating with me and turning my life around. At the end of my thanks that morning, a clear male voice told me, *'We offer soul freedom.'* And so they do. Our knowing that life never ends, our soul is invincible and remains with us throughout all lives, wherever lived, can give us the freedom to live as we choose. If we are wise, we will live good, helpful lives and thus be able to fully enjoy life in spirit.

"The third message that I always hold close to my heart, for several reasons, was their coming through one morning and telling me, *'Death no more a casket.'* How true this is! It shows that my guides, no doubt from the time I was born, knew how I'd become convinced when visiting my grand-parents at the age of six and seeing dead people in caskets, that death was a casket. They had always felt they had to convince me otherwise that life never ended.

"They were no doubt instrumental in bringing the knowledge of audio voice phenomena into my life and helped me with all the verbal contacts that have been received over the years, as well as in other areas. I owe them a debt that I can never re-pay completely. The best I can do, besides knowing that God does exist and that death never happens, is to try to help others as much as possible. Eventually, I will know what grade I've earned. I hope it will give me some joy."

The history of the field of EVP and ITC is short; but already, there are important pioneers who have worked hard to open the door for future generations. The very thought that these phenomena might exist is absurd when considered in the light of modern science. Yet, these pioneers have persisted, and in doing so, have laid the solid foundation of rational understanding which we enjoy today. Sarah Estep will undoubtedly go into the history books as a modern pioneer in our field. The EVP and ITC community owe Sarah our respect and our gratitude for the work she has done in our behalf. She has mentored many people who sought to record these voices for themselves.

Chapter 3

Lisa and Tom Butler and the AA-EVP

In this chapter, we would like to bring the AA-EVP history up to date by explaining how we came to assume leadership of the Association, and in doing so, tell you a little about who we are.

We met in Sacramento, California, in 1982. Both of us had previously been married and we were comfortable in our single lifestyles. Today, it is easy to look back and say, "Of course, we were brought together so that we could shepherd the continuation of the AA-EVP." But at the time, the quick attraction and decision to marry after just two weeks surprised us both. There was a kind of "knowing" between us from the very beginning, and we entered into our lives together without hesitation. From our view, the thought that we could as easily have not met, or that we could have chosen our single lives over marriage, are unthinkable.

Both of us had previously been interested in various aspects on New Age thought. Tom had studied The Silva Method,[79] Builders of the Adytum,[80] Rosicrucians[81] and had even ran a personal improvement center at one time. As Tom puts it, "Collectively, these systems of belief teach a person to have a sense of order in reality and to know that a person can learn to be in agreement with that order. My early experience with these systems of belief set the tone for all of what would come in my life." Tom is quick to admit that knowing things about reality is a far cry from knowing how to apply those things, so he is always seeking ways to live what he has learned.

Lisa had read many books in her youth about personal improvement and the power of belief and had learned to use affirmations to bring things she desired into her life. In fact, she was so good at creating her reality that it sometimes spooked her. She quickly learned that it was possible to bring things into her life that she did not want if she was not careful in her visualization. All of her experiences did, however, bring learning, which is of course our purpose for being here. Lisa used to call it, "witching things up," and naturally, she attributes our meeting to this ability. In truth, though, we both know that it was much more than affirmations that brought us together.

The early days of our marriage were filled with the typical trials and tribulations of a couple without much money. Vacations were often spent sleeping in the back of a station wagon at camp sites. One fond memory is of a meteor shower we saw while sitting on the tail gate of our station wagon, near Mount Hood, in Oregon. Then there is the Wild Mountain Cat, the park manager's cat that jumped on our car in the middle of the night. We knew for sure that we were about to be abducted by aliens.

Both of us were seekers, and naturally, we quickly rekindled one another's interest in metaphysics. One of the more interesting authors we found was Paul Twitchell, and it was while we were reading some of his books that Lisa had an important experience. She had received a call that her father was ill, in the hospital, and not expected to live. Her mother was upset and told her to catch the next plane home.

Her mind was spinning as she tried to figure out what to do next. She decided that she needed to meditate to calm herself down. What happened next had not been experienced by her before. She experienced a vision of Paul Twitchell and his spiritual teacher, Rebazar Tarzs. In her vision, she came upon the two men sitting on a log with her father and was told that it was not his time yet. The experience was not like a normal meditation but more of an out of body experience, which is what Paul tried to teach. The meditation experience did indeed turn out to be true and it was not until several years later that her father made his transition.

California weather and outdoor life were central to our lifestyle, as was a large garden and also a green house for the many lovely orchids we had collected. Nature photography became an important hobby for us and we always took our cameras when we traveled. Then in 1987, Tom came home from work and said that his job had been moved to Kansas City. He said that he would have to let the company move us, or he would have to find other work in Sacramento.

Culture Shock

If you know anything about West Coast living, you will understand when we say that the move to Kansas brought considerable culture shock for us. As it turned out, we ended up working for the same company, and financially, we were doing just fine. However, it seemed that working for a corporation drained our spirit and living in Kansas did little to help us replace that loss. It took us nearly three years to finally unpack all of our boxes and hang our pictures on the walls. Almost from the beginning of our stay in Kansas, we were looking for a way to move back to the West Coast.

Although we did not adapt well to Kansas, we realize now that it was exactly where we needed to be. There was not much time in our lives due to the long hours we worked. But because of the long winters and humid summers, whatever spare time we had was focused on metaphysical pursuits, rather than outdoor activities. We discovered EVP and Tom began writing a book about metaphysics.

As with so many other people, our introduction to EVP came with Sarah Estep's book, *Voices of Eternity*. Even with our background in metaphysical concepts, we found the idea of being able to record the voices of discarnate people outlandish, even improbable. Nevertheless, Sarah made a good case for EVP and we could see that finding out for ourselves would be easy. After all, what if she was right? As it turned out, it did not take long at all until we recorded EVP messages on our old reel-to-reel tape recorder.

It is difficult to describe the feeling of wonder we felt when we knew that we had recorded a voice for which there was no mundane explanation. The truth of what we had done was shocking! The paranormal voices became part of our lives, as we recorded nightly, and then spent hours deciphering what had been recorded. Interestingly, we quickly learned that we could not talk about the amazing phenomena to anyone that we knew. The reactions and strange expressions always came quickly, making it clear that our friends thought we were delusional. Wanting very much to know that we were not, we joined Sarah's AA-EVP and were relieved to find a group of fellow experimenters who were receiving and recording the voices of those in other worlds.

From EVP messages that we received, we quickly satisfied ourselves that the voices were not stray sounds in the environment or stray radio broadcasts. In fact, with deceased relatives giving us messages on our audio tape, we came to realize that EVP was the most important proof of survival that we had ever experienced.

The AA-EVP NewsJournals written by Sarah were a wonderful gold mine of ideas, and in an attempt to improve the quality of the voices, we tried many different devices and sound sources that we learned about from the NewsJournal. The NewsJournal also came to be an important source of assurance that other people in the world understood the importance of the voices and were working to bring knowledge of EVP to the public.

When we took a vacation from our demanding jobs, it was always to a place that Lisa had found that offered personal development, such as the Light Institute[5] about which Shirley Maclaine wrote. For our spiritual growth, we also attended classes at the Monroe Institute[4] and Delphi University.[6] These classes also helped balance our logical, hard driving and aggressive work styles.

Along with our EVP work, we found a local Reiki[83] group to learn and practice spiritual healing. This was our first introduction to spiritual healing and we eventually became Reiki Masters. Some of the other classes included a variety of different healing methods, and we could see that all of these come from the same source. Healing methods may have different names but they are all spiritual healing and they all provide a profound demonstration that we are spiritual beings.

Naturally, we really enjoyed the personal development classes that we had attended. Soon, we found that EVP was not the only phenomena in our lives. There is nothing quite as mentally stimulating as that moment when something happens that cannot be explained with known physical principles. For instance, we were beginning to have prophetic dreams, and occasionally see lights that should not be there. Such experiences drove home the reality of our larger Self faster than any lecture, class or well-written book. The lesson was clear that there is much more to our reality than most people had thought.

Our spiritual growth served to make us want even more to escape the corporate world. Both of us were in positions that required us to interface with people in ways that we would not in our personal life. In fact, our corporate lives were quickly becoming radically different from our increasingly expanded view of reality and what was really important in life. We would return from our classes happy, with new knowledge and our minds at peace, only to have our tranquility shatter by our pressure-cooker corporate lives. Yet, neither of us knew much about life outside of the corporation and it was difficult for us to imagine our alternatives.

After writing out goals of what we wanted our life to be like, we meditated. Remembering her uncanny success using affirmations to change her work situation against what seemed like insurmountable odds, Lisa returned to working with affirmations in an effort to create a more spiritual environment for the two of us. However, we were both educated, logical people working for the same company. On more occasions than we could count, one of us would come home from work threatening to quit, only to be talked out of it by the other. After all, we had good paying jobs and we were not sure of what we would do if we quit, but there was a hole in our lives.

Finding the Courage to Drop Out

Delphi University is a training center founded by Patricia Hayes and Marshall Smith that is devoted to healing, mediumship and metaphysical studies. Our classes at Delphi began in 1995 and continued into the next year. The classes were intense and sometimes hard on our egos, but we noticed positive changes in our worldview with the very first sessions. The week-long classes involved nearly total immersion into on-campus living, from dawn to late at night, in which we learned about mediumship, healing and teaching. Even with all of our study in metaphysics and work in EVP, it had simply not occurred to us that we could have the abilities that we had so often read about in others. In fact, EVP had appealed to us because we felt that it was something outside of us, and that we had no part in the creation of the voices. After the classes, we had little choice but to accept that our mediumship played a part in our success with EVP.

The third or fourth night of our first week at Delphi, Lisa stopped to speak with some of the other women who were attending the same class, while Tom continued on to bed. It was very late when Lisa quietly returned to the room where Tom was already sleeping. She was still trying to fall to sleep when she smelled cigar or cigarette smoke, and then heard a man speak to her. In Lisa's words, "It was as if the voice was in the room and outside of me, but it was not waking Tom so I guess it had to have been in my head. The man said that he was Arthur Ford, the mentor of Patricia Hayes and the Godfather to her children. In fact, we were sleeping on campus in the Arthur Ford Hall, and Marshall Smith, Patricia's husband, conducted channeling sessions for the students in which Arthur spoke through him.

"The voice of Arthur Ford gave me three messages. The first was a personal message for me. The second was that Tom and I would be running an organization that would help others learn that death is not the end. The last message was to tell Patricia to get business cards. Even as he spoke, I was wondering how my mind could be producing such incredible happenings. I decided that I was simply hallucinating because I had not been getting enough sleep. Satisfied with that explanation, I rolled over for some sleep. But, the voice was having none of that. The voice, calling himself "Arthur," told me to write the messages down, because otherwise, I would forget them. In my head, I told him I could not do that because turning on the light would wake Tom. He was insistent and also humorous. 'You write them down,' he said, 'and I'll let you go to sleep.' Reluctantly, I turned the light on and wrote the messages down. The voice went away and miraculously, Tom did not wake up.

"I read over the messages the next morning. They were pretty far fetched. To me, the very thought of Tom and I running any kind of organization like that sounded like wishful thinking since we were working ten to twelve hours a day for the company we were with and there was little time for anything else. And the last message proved that I had been hallucinating. Patricia Hayes had a successful school and was a well-known medium. She most certainly had business cards. What a ridiculous message. But then, I would never know what had happened for sure unless I asked Patricia if she had business cards.

"That afternoon, Patricia was photocopying something for me and I had a minute alone with her. I gathered my nerve, and feeling more than a little foolish, asked her if she had business cards. I will never forget her reply. 'Arthur has been pestering me about those darn business cards. I just never seem to get the time to get it done.'" The classes showed us that everyone has a psychic ability and with work it could be brought out.

It was when we were boarding an airplane to come home from one of these week-long classes when we both decided to quit work. We were often high on energy right after the classes. During the wait in the airport for our flight home, Lisa said that she was seeing energy around people that was much like the human aura. In awe of this new seeing ability, we were laughing and enjoying the moment. Somewhere amongst the laughter, we came to know that things had changed in our lives.

One of the outcomes we believe people hope for when they attend personal improvement courses is that they will undergo some form of life-changing transformation that will help them escape their mundane lives as they come to fulfill some divine purpose. People seek significance in their lives, and we are no different. For us, Delphi did offer life-changing insights that eventually led us to find the courage to make a profound change in our lives.

The very next day after we returned home, we put the house on the market. It was the beginning of winter, which you will know is not a very logical time of year to sell a house, if you have spent a winter in Kansas. Nevertheless, we persisted, and began by either selling or giving away everything we owned that would not fit in a trailer and a ten by twenty foot storage compartment. Our intention was to use the money to buy a fifth-wheel trailer and spend the following summer in Alaska. After that, we planned to return to the West Coast where we would stop in Portland, or Seattle, and find jobs. Neither of us talked the other out of executing this grand scheme.

In March of 1996, we had our fifth-wheel trailer and a huge truck to tow it, and had begun a journey that would introduce us to some beautiful soul-filling country. It is surprising how little fits into a home on wheels. Our EVP recording gear took up precious closet space, but we gladly sacrificed the space. Would the voices be able to be recorded on this road trip? As it turned out, travel was no barrier for them, but we learned that the voices came in better at different locations. Were these places ley lines, or so-called power spots? Could earth energy be yet another equation in obtaining better quality voices?

From Kansas, we went south to Albuquerque, New Mexico, since that is where we chose to rent a storage compartment. Lisa had been raised there and her mother still lived in the valley. From there, we slowly traveled north, taking our time so that we would reach the southern end of the Canada-Alaska highway in warm weather. Along the way, we experienced what we remember as a never-ending spring. The buds on the trees had not yet begun to swell in Kansas when we left. There was just a hint of green in New Mexico, and it was spring for the next few months as we traveled north, staying just south of the frost line. This is a huge country with vast beauty throughout.

Along the way we learned that we could make it without going back to work. If you don't have much of a place to put things, you don't buy much. So, when we returned from the beauty of Alaska, we passed right through Seattle and Portland, confident that proper budgeting would allow us to avoid returning to the corporate world.

Our plan was to spend the winter along the Northern California coast, but before settling into a winter trailer park, we stopped in Mount Shasta, California, to check on a piece of property that we had once intended to make our retirement home. Mount Shasta is a beautiful place that is dominated by a volcanic mountain whose white peak is visible nearly to Sacramento. The town is also awash in young New Agers who make a pilgrimage there to experience the spiritual energy said to be around the mountain. It had always been our belief that this New Age energy was a lot of hype, but when Lisa woke up after our very first night there; she said that she had had a dream. A being had come to her in the dream and told her that we should not stay on the coast but should spend our winter in Arizona. Experience had taught us to allow ourselves to be guided by such encounters with Spirit. How can you argue with a dream?

It was September when Lisa had the dream, and knowing that thousands of "Snow Birds" would have already planned their journey south for the winter, we immediately began making calls to trailer parks in an effort to find a place to stay in Phoenix, Arizona, for the winter. There were no spaces available in any of the more desirable parks, and out of desperation, we began calling what we considered the "C" list until we found one with a space that was big enough for our fifth-wheel.

In November of 1996, we arrived in Phoenix to find the people at this trailer park to be very nice, but it was at the end of a busy airport runway, and small planes were constantly taking off and landing. However, we soon learned that we were once again where we were supposed to be. Metropolitan Phoenix is forty or fifty miles across by road, yet the trailer park was only blocks from the Church of the Living Spirit, Spiritualist Church in Glendale, Arizona.[8] We attended their service that first weekend and met the pastors, the Reverends Gene and Sandy Pfortmiller. Sandy commented that she felt like she knew us from before, and we felt the same. Since then, we have been very close to Sandy and Gene.

One important thing that we learned is that when we mentioned we worked with EVP, Sandy and Gene not only knew what we were talking about, but also believed that the communication came from those in spirit. It was something of a shock for us to learn that these Spiritualists not only knew about EVP, they even had a brochure about EVP on their church table that had been written by the Reverend Bernard Baker. Bernard was the director of the Department on Phenomenal Evidence within the National Spiritualist Association of Churches (NSAC). His name was familiar to us, as he was also a member of the AA-EVP.

One of the more important services provided by many of the Spiritualist churches are classes that include instruction in spiritual healing, mediumship and Spiritualist concepts. It has become our practice to recommend such classes to people who have questions about survival and other metaphysical concepts, as we have found most Spiritualist instructors to have a strong desire to help others develop their potential.

We had met a wonderful group of people in Phoenix. That, along with watching the news and seeing that the trailer park we were going to stay in on the California coast had been partially washed away during an El Niño storm season, led us to believe that we had indeed been guided to Arizona.

The NSAC, of which Sandy and Gene's church was a member, supported EVP as proof of survival. NSAC Spiritualists have no dogma in the usual sense, but they do have what they call a "Declaration of Principles." There are nine of them. Principle Five states, "We affirm that communication with the so-called dead is a fact, scientifically proven by the phenomena of Spiritualism." If you know anything about the way we approach nonphysical phenomena, you will know that it was easy to embrace the Fifth Principle. In the same way, NSAC Spiritualists have embraced our approach to phenomena.

It was another winter and we were once again in Arizona when we received a telephone call informing us that Lisa's mother was near the time of her transition. Lisa flew to New Mexico to be with her mother who made her transition on Halloween of 1998. Her mother had left a small inheritance and with this, and a careful budget, we were hopeful that we would never need to return to the corporate world with its accompanying stress.

Phenomena

One of the things we really missed while on the road, was the intense physical phenomena that often happened during our immersion in weeklong development classes. Oh, sure, we had some experiences. For instance, while gold mining in the Arizona Desert, we heard a few voices that may have been from outside of our dimension. Some of the old-timers we mined with knew of these voices and refused to camp in that mining area. Lisa also saw an old miner at a different dig site who was probably a ghost. Again, our friends, who were familiar with the area, confirmed that they knew of others who had seen the old man who looked very much alive; but otherwise, could not be found in the area.

Wanting to see if we could pick up EVP in these areas, we purchased an inexpensive cassette recorder. But it proved to be a waste of time, as we picked up very little EVP in areas that we knew to have haunting activity. Looking back on this, we know it had to be that cassette recorder, because since then, we have collected amazing EVP in the field with the IC recorder that we have now.

The second winter that we were in Phoenix, Lisa subscribed to the Morris Pratt *Correspondence Course in Modern Spiritualism.*[10] After seeing that she was determined to finish the course, Tom signed up as well. The Morris Pratt Institute is affiliated with the NSAC and completion of the course is a prerequisite for all of the church's certifications. It took several years of training and study, but we became NSAC ordained ministers with credentials in healing, mediumship and teaching. In August 2001, we were appointed the Co-Directors of the NSAC, Department of Phenomenal Evidence (DPE).

We should make a comment here about the synergy between Spiritualism and the AA-EVP. While it is true that much of our work in behalf of either the DPE or the AA-EVP also applies to the other, we keep the two interests separate. EVP is not about religion and we have taken pains to keep religious overtones away from the AA-EVP. In fact, reading this book will probably provide the first inkling for many of our AA-EVP members that we are Spiritualists.

The AA-EVP

In May of 2000, Sarah Estep asked us if we could take over the AA-EVP. Our first realization about taking over the Association was that it would mean another major life change. Perhaps the road trip was over and we were ready to go back to work, doing something that might make a small difference.

A little stunned, and torn between wanting to help Sarah and the recognition that we were not Sarah, we agreed to wait at least a day before responding. As Lisa recalls, "That night I had many beings appear to me in a dream; a very vivid dream. They said, 'Tom thinks you can do this and Sarah thinks you can do this. You can do this and we will help.' I told Tom about my dream when we woke that morning. We called Sarah right away." Sarah told us that she had worried over what to do about the AA-EVP and felt that the Association was important to many people. She told us that she had felt guided to approach us about taking the AA-EVP over. This is amazing on both our parts, as the three of us had never met in person.

Hoping that we knew what we were doing, we agreed then and there to do our very best to continue the Association. Sarah said that she would send a letter of introduction to the members, along with money for the remaining value of their membership. She sent us the membership list, and as soon as it arrived, we mailed our own introductory letter to the members, along with a new membership form. Sarah had been steward of the AA-EVP for many years and she had a long list of loyal members. Would they rejoin if she was no longer its leader? Even though we had been experimenting with EVP for over twelve years, at the time, we were unknown in the EVP community. Knowing this gave us good reason to worry that perhaps just one or two people would rejoin. What would we do then?

Our mail was delivered to us via a forwarding service that added a week to the usual delivery time for mail service. So, we did not expect any membership forms back in the mail that first week. To our great relief, there were twenty-eight. Many of the letters kindly thanked us for carrying on Sarah's work and described how important the AA-EVP had been to them. Encouraged, we decided that we could give the Association a fair shot even if we only ended up with fifty members. The next bag of mail brought eighty membership renewals! The AA-EVP was alive and well.

A House for the AA-EVP

Since we were in Reno, Nevada, and had the fifth-wheel at a year-round park, we actually had a telephone line. However, after our agreement to assume leadership of the AA-EVP, came the realization that our 19.2 Kbps wireless, Internet modem and single telephone line was not enough to support the work we had ahead of us. There was a new website to build and maintain and we knew that we wanted to open an email distribution service for members to share emails and EVP samples with one another. Obviously something had to be done, and it was then that our favorite pastime, "window shopping" for a home, turned into a serious search for an office so that we could have more than one telephone line. And yes, a place that would have room for us to eat and sleep along with a real experiment room, as well.

The small town of Reno had always given us good feelings when we visited. The view of the mountains and the big sky that was often filled with lenticular (lens) clouds were our favorite. It took no more than twenty minutes to drive from one end of town to the other; it had an airport, a shopping center, plenty of grocery stores—everything we needed. That is one of the reasons that we spent most of each summer there the last few years we were on the road.

Interestingly, we had looked at houses before, but felt that we could not afford them. Our financial resources were and are fixed, and the idea of committing ourselves to the cost of a house was more than a little intimidating. However, as we examined the cost of owning a home in comparison to the cost of living in a trailer, we came to realize that we could not afford not to give up our life on the road. As it turned out, living in a trailer is a perpetual financial drain, while a home always offers the possibility of appreciation.

There was a special area that we often toured when we "window shopped" for that house we would like to buy someday. The area was a hill to the north of town and above the University. The same year, and before we agreed to assume leadership of the AA-EVP, we had once again toured "the hill" and stopped by a particularly inviting house that had a "For Sale" sign in front. The place stuck in our mind because the box that held fliers about the house had fallen with Lisa's first touch. Naturally, the owner had just stepped out the front door. After the mad scramble to help the owner run down the flyers that were blowing around in the brisk wind, more than a little embarrassed, we jumped back in the car and made our escape.

The first time we looked at the price we thought we could not afford the house. The second time, the house had been reduced and was just within our affordable range for a home. To make a long story short, we ended up buying the house.

Many EVP and ITC researchers say that, when you move to a new location, it can take a while to build up a bridge to the other side. Understanding that some locations were simply better than others were for EVP, we wondered if this would be the case with our new home. The spirit voices came in on our first experiment. They were happy with the new home they helped us find!

That first year was spent taking care of the Association, trying to make more contacts in the International EVP and ITC community and conducting our own research and experiments, while also redoing bathrooms and most certainly getting rid of the wallpaper.

In October of the following year we received our first communication from Lisa's mother. It was a shock but also much needed. She had been mentally ill in her physical life and we had been concerned that she might be stuck close to the Earth Plane. The contact also provided a major healing for Lisa, as there had been unfinished issues that needed to be resolved. A year after that, we began experimenting with video ITC. We talk about those experiences in Chapter 9.

We have no idea what tomorrow will bring into our lives, but as we prepare this book, we are very happy in the roles that Sarah Estep asked us to fill. Recording and receiving EVP messages have touched our lives as well as the lives of the many wonderful people in the AA-EVP. You will be reading about many of their stories and experiences in this book. Others outside of the Association, and from many different countries, write to tell us of their experience with EVP and the messages from those in other worlds. It is our desire that this book will encourage you to conduct your own experiments with EVP and ITC. For our part, we are exceedingly grateful that we have found a way to serve those in Spirit who want to communicate with their loved ones here on earth, to let them know that there is no death and there are no dead.

Chapter 4

Some Researchers Working in EVP and ITC

There are many researchers working all over the world in EVP and ITC. However, because they speak other languages and information on them is not readily available in English, we have been unable to report on all who probably deserve to be mentioned in these pages. There are many others who are not listed here simply due to space limitations. So what follows are just a few of those who have become well-known, have been covered in the media and/or are spreading the word about these phenomena to the public.

Erland Babcock

Erland Babcock's background is electronics and photography. He was a senior electronic technician in various departments at the Massachusetts Institute of Technology (MIT) for fifteen years. In 1982, Erland's son, David, was fooling around with a tape recorder because he had heard that one could record voices of the so-called dead. David used a radio between stations as a source of background noise. After making a recording, he took the tapes to his father's home and asked him to listen to them for unusual voices. Erland could hear something, but having a background in electronics, he always blamed the sounds he heard on just the noise of the radio or on some artifact of the equipment.

His son persisted and told Erland to "really" listen. "Listen for a series of tones and then a whispering voice," he said. After several tries, Erland did hear what was being said by the voices on the tape and was flabbergasted. What he heard was his mother's voice saying, *"Hi David, where is Erlin?"* Erland's mother always called him "Erlin." "How could this be?" he thought, knowing that she was dead and buried. This experience marked the beginning of a long career of research and experimentation, as he sought to understand the nature of the voices.

In 1984, Erland read a story in the *National Enquirer* about a scientist who talked with the dead. Erland was working at a University at the time and so wrote to the scientist using the University letterhead. The scientist was George Meek. Meek eventually gave him a grant to work on EVP and ITC in the fields of possible magnetic influences on EVP, and lasers as a source for both EVP and optical ITC. As part of Meek's team, his research focus was primarily on laser technology, while Hans Heckman focused on magnetic field research.

Erland had read about the work of Klaus Schreiber who had received a picture of his discarnate daughter using video equipment. At that time, Erland was working in MIT's Audio/Video Laboratory and believed he could apply some of his expertise in Video ITC experiments. He copied Schreiber's video loop method, which is very much like the one we describe in Chapter 13, and immediately began receiving phenomenal features. Erland may be the first person to work with Video ITC in the United States.

Over the years, Erland Babcock has received many wonderful ITC images, including scenes of people, animals and landscapes. Samples of his work are included at aaevp.com.

Stefan Bion

Stefan Bion always asked questions like, "Where was our consciousness before we were born; where do we go after we leave behind our physical body and what is the meaning of this 'play' called life?" He began to read esoteric literature in his search for answers. In 1987 he met an author of one of those books and was told about Electronic Voice Phenomena. Stefan had some knowledge of electronics and saw this as an ideal field of research.

He found the address for the German EVP association, Vereins Fur Tonbandstimmenforschung (VTF), and contacted them. At a monthly meeting of one VTF group, he was able to convince himself that EVP was real. He conducted his first EVP recording as soon as he returned home. He wrote, "I took an elektret microphone with a preamplifier, which I put into a glass bottle with a wide neck for the purpose of resonance. I have no idea whether this made sense. It was approximately 1:40 a.m. and absolutely quiet. I made the first recording and analyzed it, nothing except the ticking of my alarm clock. I removed the battery from the clock. Second recording at 2:00 a.m. I listened again and suddenly, like out of the void, there was a voice which sounded like a talking budgie: *'You see, I can't do this!'* It was overwhelming!

"Next I visited a local VTF group in Cologne. Mrs. K. told me, 'We're recording using water.' I didn't understand what she meant. When we entered the 'EVP recording cellar,' I saw the mysterious equipment: Several funnel-shaped piezo speakers hung down from the ceiling, and a wire connected them to a funny-looking metal tea-can, which, after removing the cover, turned out to be a self-made microphone preamplifier. A reverberation device and a cassette deck followed. The mentioned 'water' was a fountain in the garden pond whose splashing noise penetrated quietly through the cellar window into the room. When the recording started, I was quite curious. Each one of us asked some questions, and upon playback, I noticed the peculiar sound of the recording which was probably caused by the frequency response of the piezo speakers which were 'abused' as microphones. The original splashing noise was no longer identifiable as such. Instead, a strangely croaking noise was heard which sounded somehow articulated and sometimes actually contained words and even sentences! Especially astounding was a voice which seemed to make a technical comment, *"Has of course also a sound switch—contact!"* I was really stunned. It was always said and written that EVP were very faint and only decipherable after a long time of practicing—and now there was such a clear voice which I was able to hear immediately!"

A time of experiments followed, in which Stefan tested all known recording methods. He wrote, "I soon noticed that the quantity and the quality of the voices does not depend primarily on the technique used, but that there's another 'unknown factor,' which lets the results vary in temporal periods. As frustrating as this can sometimes be—it clearly shows that EVP are *not* merely a technical-physical or a perception-psychological phenomena. I experimented with different kinds of acoustic 'raw materials' and now use a computer to produce a synthetic background noise which can be used by the 'voice entities' to form their messages. The software is free for anyone who is interested in this method." This software is known as EVPMaker[72] and can be downloaded from the Internet at no charge at:

www.stefanbion.de/evpmaker/.

Since 1997, Stefan has been the administrator of the VTF website, and since 2001, has been the editor of the *VTF-Pos* German language EVP magazine.

Stefan tells us that, "Besides my technical interest, also the philosophical and spiritual aspect of this phenomenon has always been important to me. Perhaps this research can contribute to a more humane world by making people realize that there could be something beyond our pure material existence, that life does not lead us into a dead end and that we are all embedded into something greater.

For details on contacting the German EVP organization, the VTF, please see information under Jutta Liebmann in this chapter.

Jacques Blanc-Garin

Jacques Blanc-Garin of France and his wife Monique Laage, work together in the field of ITC. He has a strong computer technology background and currently operates an information security company.

Jacques' introduction to EVP came in 1986, soon after the transition of his first wife, Annick. He received his first EVP from Annick nine months after his first recording efforts. This short message, in which he recognized Annick's voice, was, *"Je suis là"* (I am there).

Jacques began experimenting with Video ITC in 1991 with his research friend, Monique Simonet. Some of their results were published in her books about ITC. It was in 1992 that he met Monique Laage and founded the ITC association, Infinitude. The focus of Infinitude is to help people who are mourning. EVP is used to communicate with the deceased. This helps people accept the loss of a loved one and also learn they will continue to survive. Infinitude offers various forms of support for those who are grieving. This includes a website, www.infinitude.asso.fr, conferences, and a magazine.

With sixteen hundred members, Infinitude is a very active Association. Jacques has told us that, "Since the founding of Infinitude, we have provided eighty lectures, conducted more than one hundred and fifty experiments during conferences and have conducted more than four hundred and fifty EVP sessions that enabled grieving families to contact their loved ones on the other side. We have focused our efforts on grief management sessions with people in mourning, but we also assist people who need help understanding and dealing with haunting situations. To facilitate this, we have begun travelling to various provinces throughout France to meet with people who are not able to travel to our city. As part of these travels, we have visited fifteen towns since 1998, in which we have held congresses to present ITC and have participated in six television shows."

As testimony to just how active Infinitude is, Jacques has told us that they have published an ITC magazine, first known as *INF'INIT*, then *Infomonde Tci* and now an excellent trimester publication known as *Le Messager*. They have also released two audio tapes with EVP samples titled, *Voyage dans l'Au-delà et Messages de l'Au-delà*. Finally, he and his wife, Monique, have written a book about their story and their work with ITC titled, *En communion avec nos défunts, dans l'infinitude de l'amour,* edited by Les Editions du Rocher in June 2002.

Jacques Blanc-Garin closed his comments to us with, "The most important part of our work is to help people who have lost a loved one, but we also participate in research projects."

Anabela Cardoso

Anabela Cardoso is Portugal's consul general in the Rhone-Alps region of France. She has a Ph.D. in Germanic Philology and was Portugal's first female consular.

After being deeply stricken by the grief of losing a loved one, and in an effort to find out what really happened to those whom she loved and missed, Anabela and a small group of people began attempts to make contact with other realities by means of EVP in 1997. Two other members of the four-member group were also suffering from the loss of a loved one.

The background noise used for these communications was that of three radios variously tuned between station on the short wave and AM bands. The group customarily sat with a tape recorder, asked a question, and then waited in silence for a time. After which, they rewound and listened to the recording to see if an answer had come via EVP. At the end of two and a half months, the answers began coming, sometime including words in two or three different languages. Anabela's mother tongue is Portuguese but she speaks other languages as well.

The little group eventually disbanded, but Anabela continued trying to make contact with those on the other side. She had read Hideard Schäffer and Maggy Harsch-Fishbach's books on ITC, from which she learned about Timestream Station. Timestream is a group of thousands of people in spirit, many of whom are devoted to establishing communication between their world and ours. She included questions about Timestream in her attempts to make contact with the other side. On March 11, 1998, Anabela was alone at home conducting an EVP experiment when she heard her first direct radio voice. The contact on the radio from the other side was made by Carlos de Almeida of the group, Timestream. She writes that this was the most memorable day of her life and that she was in a state of shock for twenty-four hours and unable to turn the radios on for a couple of days.

Anabela Cardoso established the *ITC Journal,*[68] with the hope that it would provide an incentive for those who wanted to know more about Instrumental Transcommunication and provide a guide for others wishing to learn how to experiment with the phenomena. She is the Director and Editor of this nonprofit publication and there are many highly respected EVP and ITC researchers on the advisory board. The journal is published quarterly, and includes articles by renowned researchers and specialists in the field of EVP and ITC. For more information on the *ITC Journal* see the website:

www.eureka.ya.com/cadernostci.

Judith Chisholm

Judith Chisholm is a journalist living in London who unexpectedly lost Paul, one of her sons, in 1992. Overcome with the grief from this loss, Judith embarked on a search to learn more about what had happened to Paul after his death.

Judith met another woman whose son had also died, and was told about a weekly séance that met near her home. She began attending the sittings, and one evening, took her tape recorder and placed it on the table in the center of the circle. When she later listened to the recording she heard a woman saying, *"Judith."* This intrigued her, as she was certain that her name had not been spoken during the sitting.

After some research, Judith learned about EVP and that the voice might be an example of that type of phenomena. she began working with her other son, Victor, to record for EVP. They sat for an hour twice a week, recording and examining the resulting sound tracks. Their focus was on contacting the woman who had whispered Judith's name, or of course, any other spirit that would communicate.

The experiments continued for six months without results. Then, just as they were about to give up, they finally heard something on a recording. A man's voice said, *"I've been every week!"* After this the voices came more frequently and Paul, the son whom Judith so desperately wanted to speak with, recorded his voice in early 1994. This was eighteen months after his death.

Judith Chisholm has recorded hundreds of paranormal voices and written a book about her discoveries titled, *"Voices From Paradise."*[57] In 1996 she started The EVP and Transcommunication Society of Great Britain and Ireland in order to disseminate information about EVP. She publishes a quarterly newsletter for the Society, has promoted EVP on television and has provided many presentations and talks promoting the voices from the invisibles. You can go to www.voicesfromparadise.co.uk for more information on the EVP and Transcommunication Society of Great Britain and Ireland.

www.voicesfromparadise.co.uk

Pascal Jouini

Pascal Jouini is a French researcher who has been working in the field of ITC for over fourteen years. The exploration began with his very passionate interest in ghosts. His search into this field is purely experimental and apart from all philosophies and religions.

His website (http://perso.club-internet.fr/pjouini/menugb.htm) displays many of his ITC pictures, as well as diagrams and pictures of his ITC equipment. There are also images displayed that have been captured by other experimenters. One feature that sets Pascal's website apart from others is his online Video ITC webcam. He usually runs a Video ITC experiment on Sundays. People from around the world can visit his site and see what an experiment looks like, and even participate by capturing "screen shots" while the experiment is running.

Tina Laurent

Tina Laurent, England, had read Konstantin Raudive's book *Breatkthough*,[2] but it was not until 1981 that she heard her first voice. She wrote, "I remember the first time I listened to discarnate voices emanating from magnetic recording tape! It was 1981, in the Maryland house of Sarah Estep. For three hours I sat enthralled, listening to the voices and knowing full well that this was a day that was to change the course of my life."

When Tina returned home she borrowed a tape recorder and began attempts to communicate with any passing spirits who might hear her. She immediately heard unintelligible whispers and was certain that they should not have been there. Within two weeks, Tina heard her first intelligible voice saying, *"Tina."* Since then, she has contacted her EVP friends on a daily basis.

Tina Laurent has written several articles on EVP, provided many public presentations spreading the word about the phenomena and has appeared in numerous TV programs.

Jutta Liebmann

Jutta Liebmann, of Germany, read about Konstantin Raudive's book, *The Inaudible Made Audible* (original version of the book later translated into English as *Breakthrough*) in a fashion magazine at the hairdressers. Due to her interest, her parents bought the book as a gift for her. She read the book with a mixture of skepticism and belief.

Jutta put the book away for nearly two years, but something lead her to return to it, and it was not long before she began experimenting using a microphone and reel-to-reel tape recorder. After six weeks, she recorded her first word, *"Contact,"* and this word was repeated during her next recording session. Jutta, was a little bit frightened at first, and tried to find an EVP recording group for support.

Jutta became acquainted with Hanna Buschbeck, an EVP pioneer who was coordinating experimenters in Germany. At her first meeting with the group, she had the privilege of meeting the great pioneer Friedrich Jürgenson.

EVP experiments changed her worldview to a more spiritualistic attitude towards life. The authenticity of EVP brought Jutta to the realization that there is a spiritual world and that individuals survive bodily death. She has received reasonable answers and statements from the spiritual world in reply to her questions concerning many topics of general interest. She has told us that she has received answers "concerning the existence of non-human entities ... in the cosmic worlds, i.e. entities originating from another evolution than mankind."

Some of Jutta's friends in spirit were interested in extraterrestrial life and have given her fascinating answers during her numerous EVP recording sessions. In the mid-eighties she began recording long messages that appeared to be transmitted with an extraordinarily strong energy containing strange background frequencies. These messages sounded quite different from those that she usually received, leading her to believe that they came from a different dimension and might even be of extraterrestrial origin. She wrote, "I have never before or after recorded such long and clear messages. Therefore, I base the fact that ninety six percent of the EVP contacts originate from the spiritual worlds and the rest come from extraterrestrial and/or inter-dimensional realms of existence."

She has received thousands of EVP voices. She has recognized the voices of deceased relatives and tells us that many messages have been extremely helpful in her personal life and have helped her to avoid harm on several occasions. Her audio archives include many clear audible voices that have been presented at the twice a year Vereins Fur Tonbandstimmenforschung (VTF) conferences. Jutta has been a member of this German EVP and ITC group since it's foundation by Fidelio Koberle in 1975 and has been on the board of directors of the VTF since May 2000.

The VTF's goal is to prove the spiritual survival of bodily death. They publish a quarterly magazine. For further information on this organization you can go to their website at:

www.vtf.de.

Alexander MacRae

Alexander MacRae (Alec), Scotland, began researching EVP in 1979. As part of his study, he developed a biofeedback device designed to detect changes in the Galvanic Skin Response (GSR) of mediums. The Alpha, as he named it, included self-balancing circuitry that caused a light to change color as the GSR changed. It also generated audible tones that changed in frequency with a change in GSR.

Alec was conducting experiments on the device in 1982 when the Spirit of Serendipity intervened. He was trying to increase the sensitivity of the device and was frustrated. Deciding to take a coffee break, he sat the device on top of an old clock radio. Unknown to Alec, the radio was actually turned on but was not tuned to a radio station. He had the GSR contacts on and they were connected to the Alpha. Quite unexpectedly, he found that sounds came from the radio in response to his every movement. Even his surprise registered as a change in sound. He tried to take down some lab notes on this unexpected event but his every move caused new noise in the radio. A small tape recorder was close by and he used it to dictate his notes about his experiment so that he would not have to move the GSR contacts while writing.

When Alec reviewed the tape recording containing his lab notes, he was shocked to hear a voice that was not his own say, *"Carl Johnson."* He rewound the tape to see if there were other EVP that he had missed and there were. The first message was the voice of his father who had died eleven months before.

This serendipitous combination of events resulted in the development of the "Alpha" Device. The Alpha has gone through many modifications and Alec continues to work on improvements to the device. In March of 2003, he flew from Scotland to the United States to test the Alpha Device in a laboratory at the Institute of Noetic Sciences (IONS). On March 03, 2003, at about 11 a.m., EVP messages were recorded in an IONS laboratory, in California, using the Alpha Device. The lab was shielded from all electromagnetic radiation, such as radio waves or laser beams. It was also shielded from all acoustic waves including audible sound, infrasound or ultrasonic waves. In Alec's opinion, "There is no way known in our present science that the EVP could have been recorded in this laboratory by physical means."

For more information on Alexander MacRae, visit his website at: http://aspsite.tripod.com/.

Mark Macy

Mark Macy was raised in Windsor, Colorado where he earned degrees in journalism and electronics. He worked for two decades as a professional writer and editor on newspapers and in technical writing departments of high technology corporations. In the early 1980s, Mark was an atheist who was firmly convinced that notions of God and an afterlife were no more than wishful thinking.

In 1988, Mark was diagnosed with cancer. He read many books during his return to health, some of which opened his mind to the possibility of a higher reality. However, his skeptical mind still needed proof that the things described in books by such authors as Paramahansa Yogananda, Herman Hesse, and Peter Hayes, were real.

Mark found that evidence in a new field of research called Instrumental Transcommunication. ITC represented the research of a few individuals working on spirit communication through technology. The findings of George Meek, Maggy and Jules Harsch-Fischbach, Adolf Homes, and Fritz Malkhoff, were so amazing and so indisputable to Macy that he became immersed in the work of promoting and sharing information on ITC with the world.

As one of the founders of INIT (the International Network for Instrumental Transcommunication), Mark arranged funding for the annual meetings of its members who were scientists and ITC researchers from various countries. During the 90s, Mark published a journal that provided information on the amazing ITC results of the research team, Maggy and Jules Harsch-Fischbach, Adolf Homes and others.[62] He later published the journals *Contact* and *Transdimension,*[56] through his organization, Continuing Life Research.

In 2002, Mark formed the organization, World ITC, with ITC researcher, Rolf-Dietmar Ehrhardt. This is a nonprofit corporation for scientific and educational purposes, "…to promote decency in human relationships, to sustain resonance among ITC researchers and to forge a link with the light, ethereal realms of existence."

Mark spreads the word about ITC research through presentations, radio and television interviews, and has authored several books. Visit Rolf-Dietmar Ehrhardt and Mark's excellent World ITC website at www.worlditc.org/. It is filled with interesting articles, historical information, video clips, voice examples and information about various researchers.

Dale Palmer

Dale Palmer is not an experimenter, but he needs to be noted in this book as one of the important supporters of the scientific investigation of what he calls, "Electronic Disturbance Phenomenon" or "EDP." EDP is a term coined by long-time researcher Professor Euvaldo Cabral Jr.

Dale Palmer has founded the Noetics Institute, Inc. (NII), a not-for-profit organization with headquarters in Plainfield, Indiana. Professor Cabral has been with the NII since the end of October 2001, working in a very specific sub-area of EDP: the voices heard in the background of files containing only recorded noise. NII researchers have been working with pseudo-random noise generated by software and with noise generated by different hardware devices. According to what they report, their conclusions up to now are remarkable and they are convinced they will be able to publish their results in the mainstream of science.

Paolo Presi

Paolo Presi has been interested in ITC since 1973.[68(10)] He has written many articles for several well-known ITC and parapsychology journals. Presi has a technical background and uses the scientific method in his research whenever possible. After years of research into EVP, ITC and related phenomena, such as direct voice, his efforts have been directed into documenting the phenomenal constraints and variables which characterize these phenomena.

Presi's research has brought him to several conclusions. He has found that the psychological conditions existing in the experimenter are fundamental for the production of the phenomena and that the qualitative results are strictly dependent on the level of mediumship existing in the experimenter. He has also concluded that technical devices used in experiments are not the determining factor in receiving EVP and do not influence either the quality or quantity of what is received. He also notes that the EVP phenomena do not follow well-established physical laws. These conclusions are based on sound research conducted by him and other researchers in the field.

Presi is one of the founding members, and on the Board of Directors, of the "ll Laboratorio" (Interdisciplinary Laboratory for the Biopsychocybernetic Research.) This is an Italian Research Association, founded in Bologna in 2002, that is dedicated to scientific and technical research into paranormal phenomena in Italy. A number of professional independent researchers, such as Presi, Marcello Bacci and Daniele Gulla, are in charge of scientifically testing the authenticity of the paranormal phenomena that takes place with the specific task of validating the authentic phenomena and exposing the fraudulent or improvised ones.

Presi, along with Daniele Gulla and Michele Dinicastro, manage an International Program in the Association for researching into the possible electroacoustical anomalies characterizing the paranormal voices. This investigation is using the most sophisticated professional software for voice electroacoustical analyses. The website for ll Laboratorio is:

www.laboratorio.too.it/.

Sonia Rinaldi

Brazilian researcher Sonia Rinaldi, began her research into EVP in 1988. She had read Jürgenson's book and other such literature on EVP in the 70s. Then in 1988, in a session with a trance channel at the Brazilian Institute of Psicobiofisics Researches (IBPP), she met Dr. Hernani. It was suggested that they should begin experiments with ITC, and they began experiments that same week. At the time, there was no literature that provided details about how to conduct ITC experiments, and they were forced to improvise. It took almost three years for the group of researchers to hear the first paranormal words. They later discovered that the long delay in their success was because there was no "station" on the other side at the time that was interested in contacting Brazil. Finally, they had news of the foundation of a group of spirit entities who were interested in developing contacts with them, and the communication with those in spirit began. Sonia told us that, "There is now a Station in the Beyond especially destined to contact Brazil, and which has now expanded to other countries of the Americas, beginning experiments in ITC is now much easier than at the time I began."

Sonia has written several books on her work and founded the ANT (AssociaçÃo Nacional de Trancomunicadores), the Brazilian ITC association in 1990. Today, this association has 1,700 members from all around Brazil. The ANT website is:

www.geocities.com/ant-tci/

Sonia tells us that those on the other side are using ITC to give people information in a very concrete way that spirit survives after the death of the physical body. This has enormous implications; for example, grief management. She has been helping parents whose children have died by making phone calls to the Beyond so that the parents might communicate with their loved ones. A transcript of one of her telephone calls is provided in Chapter 8.

Siyoh Tomiyama

Siyoh Tomiyama of Japan is a composer and guitar player. From a very young age, and as a result of a possible out of body experience, he has always wanted to know why we live and the purpose for our lives. He read many books and at the age of twenty became convinced of an afterlife. In 1997 he translated *Conversations Beyond the Light,* by Dr. Pat Kubis and Mark Macy, and introduced it to the Japanese people. He has also introduced to the Japanese people the research of the Scole Group in England, Victor Zammit in Australia and others research through the Internet and magazines. His most recent article, "The logical structure of the afterlife," was published in 2002.

Siyoh conducts ITC experiments once a month with a small group of people who have formed ITC Japan. After a deep breathing exercise they tape for about three minutes using the video feed back loop method.

Siyoh established ITC Japan and maintains a Japanese language website at:

www.spiritmusic.net.

Chapter 5

Examples of EVP

What do the messages from the other side say? In the next two chapters, we want to give you examples of what EVP messages are like. The paranormal voices can come through in some very surprising ways and not just on a particular person's tape recorder.

Over ninety percent of the reports on EVP presented here come from past or current members of the AA-EVP. Each quarterly News-Journal includes reports from members about interesting and surprising examples of EVP messages they have collected.

At the time of this writing, the NewsJournals and other historical documents are being made available to AA-EVP members in an online document archive. The archive is intended to become a valuable resource which should help people learn from the past so that they may be more prepared to work in this field.

Many of the individuals, who have heard about electronic voice phenomena, join the Association and then conduct their own experiments with the hope of reaching a loved one. Just as some people seek the services of a medium to find assurance that their loved one still lives and is all right, many pick up a recording device for the same reason. One of the important benefits of an EVP message from a loved one is how extremely evidential it can be. There is no other evidence like it.

The examples provided in the following pages will give you an idea as to the nature of EVP and of what is possible in your own recording. In some of the examples, we have used only the person's initials or first name in order to protect his or her privacy.

Hearing from Loved Ones Through EVP

One of the more dedicated original members of the AA-EVP is the amazing researcher, Clara Laughlin. In the field of mediums, some have risen to the top because of the consistent evidential information they receive. Clara Laughlin has done this through her work with a tape recorder.

Clara read a book on EVP in 1959, but it wasn't until much later that she became involved. In 1982 she, along with Mercedes Shepanek and Betty Evans, had the good fortune to meet Sarah Estep who gave them a demonstration on how to record EVP. In late July of that same year Clara began her experiments and received her first voice, *"Hello,"* on August 4. Her spirit team had made immediate contact! Clara's original spirit team consisted of her husband Tom and Dr. William Callie. They were later joined by a friend's husband, her sister Elsa, and finally by Mercedes and Betty.

Clara's husband, Tom, was in the hospital undergoing tests in 1981. He called her and told her to come to pick him up, as they were ready to release him. When she arrived, he was not in the room and she quickly found that he was in the Intensive Care Unit (ICU). When she reached the ICU, she learned that Tom had had a heart attack and that several doctors were gathered around his bed attending him. She heard them say, "Try again," and shortly thereafter, they came out of his room, many with tears in their eyes with the news of his passing. Tom had been connected with the hospital and many of these people knew him well.

One of Clara's first EVP messages came from Tom. The message was quite long for an EVP message and it was also extremely evidential. Tom's EVP said, *"After I died at Walter Reed, I awakened. I looked around for you and B. and you were not there. Then they took me to Pathology where they did research on me* (an autopsy). *You think of our beautiful life together. Don't look back. Nothing has changed. It isn't over yet."*

In a recording on May 2, 1983, Clara received these EVP messages from Tom, *"I was sick on earth. No more. No longer sick."* On May 3, she again recorded messages from him saying, *"Clara I have to say this. I am living."* On May 7, he came through again saying, *"Clara ... take this serious. I continue to live. This is the truth."*

Clara soon realized that there was one particular person on her recordings who was always there and who brought the people to talk to her when she asked for them. He called himself Dr. William Callie, Bill, or Doc Callie. It was Callie that first brought Tom to talk to Clara. Of course Clara wanted to know more about him and asked him if he had known Tom before. He replied, *"Tom was in trouble and I came to help."* This had been the time that Tom was in the hospital and had made his transition. Dr. Callie is now a close friend of both Tom, in the spiritual realms, and Clara in the physical.

Clara also has received many EVP messages from two other charter members of the AA-EVP, Mercedes Shepanek and Betty Evans. Both are now on the other side but continue their interest in proving that we live after death.

In 1991, another AA-EVP member, Walter, crossed to the other side. Both Clara and Mercedes, now also on the other side, had been friends with Walter. After Clara heard about Walter's transition, she made a recording to ask Mercedes about Walter. She was careful not to mention his name, but asked Mercedes if she'd had a recent transition over there from, "One of us." Mercedes voice replied back on the recording, *"Walter. No kidding!"*

Clara's sister, Elsa, finally crossed to the other side after a long illness. Clara conducted a recording and at the time did not know that Elsa had passed, but on the recording her spirit friends told her that Elsa had died and now was with her friends and loved ones. Clara made a call to relatives and confirmed Elsa's transition. The next day she conducted another recording and Elsa came through saying, *"It's a miracle, Elsa your sister."*

=====0=====

Martha Copeland's first EVP messages were wonderful confirmations of survival. Her daughter, Cathy, was best friends with Martha's niece, Rachel. Rachel and Cathy had been in an automobile accident in which neither was hurt, but at that time they made a pact. If one of them died, that person would try to reach the living one, through the computer. Just weeks after this, Cathy was killed in another automobile accident in which she was once again the passenger.

Rachel was upset because she missed Cathy and one night demonstrated this by yelling and kicking things around. She shouted, "Cathy, you left me alone. You said you would come back and contact me!" After a while, she turned on her computer and made a recording. It was in that recording that she heard Cathy say, *"I'm still here."* A young man's voice is then heard saying, *"How do you know they can hear?"*

After hearing Rachel's recording, Martha sat down at the computer and tried for many hours to record Cathy's voice. Frustrated by her lack of success, she began crying. Cathy's voice is heard saying, *"Mama I'm right here!"*

Soon after these first messages, Martha took Cathy's cell phone to the phone company to find the voicemail password for the phone so that she could check for possible messages that may not have been retrieved. She wrote, "I was placed alone in a small room with the phone to contact the customer service department. While on hold, Cathy's cell phone rang and then went into voicemail right away. I had to wait four hours after my call to customer service to access her voicemail. The only message was the call that I received while in that little room. It was a female voice saying, *'I know Cathy.'* I tried to return the call but it was not a working number."

Martha has also used a portable tape recorder in Cathy's bedroom and picked up her daughter's voice calling for her dogs, saying, *"Shishi, Grete."* Both dogs are with Cathy on the other side and the recording was made on the anniversary of Shishi's death.

Recently, Martha conducted a recording session on her computer with her mother, father (a non-believer), and sister. Martha's mother asked Cathy if she had been a good girl. Martha's father was the first one to hear Martha's deceased uncle's voice reply, *"You're damn right!"*

Martha wrote, "I bet my uncle got a kick out of seeing the shocked expression on my father's face. They were always playing jokes on each other when he was alive. I think he is still doing this from the other side!"

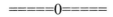

Norman Marsh, who lived in England, had lost his wife three years before he had learned about EVP. He decided to try to contact his wife through the tape recorder, but after months of trying, he still had not succeeded.

Norman wrote to the well-known English EVP researcher, Gilbert Bonner, and asked him for suggestions. He followed those instructions and used background noise as suggested by Bonner. Norman recorded a German newscaster and played a tape of this as a noise source so that those on the other side would have something to modulate to form their voices. Using this technique, he was startled to hear a woman's strong voice say, *"Norman it's Marjorie, it's true, it's true, it's true."*

Three days later, Norman sat down to listen to the message again and found no trace of the voice. The message had been totally erased. He wrote, "I had waited so long and then this happened. I was miserable but decided that I would carry on." He made another recording, again using the German language broadcast for background noise. As he was making the new recording he was surprised to hear the recording of his wife's voice on the German language tape. The recording that had disappeared from his experimental tape had been somehow transferred to the background noise tape. The tape player had been playing the German broadcast and had not been in the record mode!

Norman wrote that his wife had believed in survival by faith, but had been hesitant to accept that communication was possible. He had no doubt that the words, *"It's true, it's true, it's true,"* referred to communication through EVP.

United Kingdom researcher, Tina Laurent, knew Marsh and has said that he was a very gentle soul. She also said that she still gets goose bumps on her arms when she thinks about the first time Marsh played Marjorie's voice for her. She wrote, "I've never seen a happier man and for a few months before he died, I had some messages myself from Margorie, the same voice, saying his name and asking me to take care of him."

Art Counts cared for his sister, Lill, during a long terminal illness. Most of his taping had been directed toward contacting her after her transition. He succeeded in receiving messages from her as well as other spirit entities. One day, he heard her voice on his tape say, *"Breakthrough."* Six days after recording this voice, he was listening to some music and had almost dozed off when he heard a voice say, *"Art."* He wrote, "I sat bolt upright, wide awake, and there before me was Lill, plain as day. Not altogether solid, but not hazy either. She was looking at me with that same sweet, so gentle smile, her appearance unchanged." Art knew that her appearance was what the *"Breakthrough"* message had meant.

=====0=====

Sarah Estep has taped messages indicating that they have regular doctors and nurses in spirit who help those who have just crossed over. Her Auntie Esther came through to her many times the first couple of weeks after her death. Sarah wrote, "The first time or two, another spirit spoke for her, then Auntie Esther began speaking. The loudest and clearest message from her was, *"This is my healing except now I'm going to rush it."* This was very typical of her, and I suspect she'd quickly learned while undergoing healing, just what she could do, once she was returned to good health."

Some People Can Get Through and Others Can Not

Our own experience with getting in touch with loved ones has been a learning experience. Lisa was very close to her father, Art Zerwer, and very much wanted to hear from him. He was an electrical engineer and Lisa figured that EVP would be something that he might try to use to get in touch with us.

Our first contact with Art came about a month after his death. Lisa had a very vivid dream in which he made contact, expressed his love for her and let it be known that he was well and alive. Two more vivid dreams followed within two weeks. He showed her what he was working on and where he was living. He showed her his apartment in what looked like a quaint foreign town. The town was full of activity and her father even took her to see a movie at a theater. She could not bring back what the movie had been about, no matter how hard she tried to remember. He also told her that there was a place out in the country to which he often traveled.

These dreams were so real and so vivid that Lisa was convinced by the third dream that she had indeed been with her father; however, after this the dreams stopped. Lisa continued to try to reach Art through EVP with interesting results. Instead of her father coming in through EVP, his mother, Lisa's grandmother Nettie, began communicating. The first message received was, *"Arthur is fine and with us, Nettie."*

Lisa thanked her grandmother the next day and asked her how she was and what she was doing. The message, *"Zerwer people have house on lake"* and *"Busy lives but visit here."* Later, messages regarding the house indicated that it was in the country and we came to understand that this was where the Zerwer family would gather and visit with each other, much as we would go to our grandparent's house on holidays when they were alive.

The interesting thing about Nettie was that this was one of the last people that we thought would communicate. She had been active with a very orthodox church in her earth life, and had told the rest of the family that they were going to end up in Hell unless they joined that particular religion. Speaking to the dead was also a ticket to eternal damnation. Yet, here she was with the family updates.

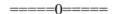

Carol Barron-Karajohn had been trying for over a year to reach a dear friend who had passed to the other side. The friend knew about Carol's taping and Carol felt that the friend would come through on the recorder, but she had not.

Then her daughter called and told her of a vivid dream that she had about a friend, Bonnie. She did not understand it but felt that Carol should call Bonnie's mother and relay the events of the dream to her. Carol did this and found that the dream had significance for Bonnie's mother, who was extremely grateful.

Shortly after this, Carol was making a recording and received a man's voice saying, *"Bonnie loves you."* Bonnie was able to get through to Carol's daughter in a dream and yet was unable to come through on tape. Her message had to be relayed by another person.

Many attempts have been made by various people and groups to establish a code word or phrase that can be given in an EVP after one makes his or her transition, so that the continuation, and therefore survival, can be proven. Sarah Estep wrote about a world renowned parapsychologist who had participated in a survival project where numerals were given that were supposed to open a combination lock. Sarah worked to contact this man and felt that she had succeeded. He provided her with evidential information but no lock number. Those who devised the survival project would only accept the lock number as proof. Sarah pleaded with the deceased man, "Give me your lock number so I can let Dr. X know I've heard from you."

Her requests were ignored or answered with, *"No."* Finally, after a week, he answered in a frustrated tone, *"I can't remember."* *"I didn't think I'd forget." "It was always with me."* Several months later the doctor's failure to communicate the lock code was mentioned in a professional psi magazine. The author wrote, "Dr. Y, always carried his number with him in his wallet."

As shown in the above story, when we cross to the other side, we may not remember things that people in the physical think that we should. When Sarah and Clara Laughlin first contacted their husbands, they found this out. Sarah reported to AA-EVP members that when she asked her late husband, Charlie, whom he first saw when he stepped through death's door, he replied, *"I don't remember. Things come back."*

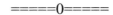

Clara Laughlin wrote to Sarah and told her that she had recorded some messages that were very similar, when she first contacted her husband, Tom. Clara mentioned an unusual experience that she thought Tom would never forget. He replied on tape, *"I don't remember. When one goes on, one looks back."* On another occasion, she mentioned a couple that they had been close to. He replied, *"I don't know these people."*

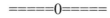

The German EVP association, VTF,[53] conducted an EVP session during a conference in Fulda, Germany. The group called on a person who had recently made his transition. Instead, another recently deceased person, Mrs. Nestler, who regularly came to the Fulda conferences, came through saying (translated from German), *"Of course, I'm present, Nestler, I was here so often."*

Precognitive EVP

Many experimenters have collected messages that correctly predicted a future event or gave them information that is unknown to them but was later verified.

Jacques Blanc-Garin and some Infinitude members had planned a weekend meeting. The first day would focus on ITC experimentation and the next day they would all go to a lovely Basilica, called Montligeon, where they would pray for the deceased. During the first day, the group made a recording and a mother whose son had died two years before asked, "What are we going to do tomorrow?" Her son answered, *"Tomorrow ... To Montligeon ... To dance."* They all thought that this was a curious answer as one prays in a church but usually does not dance. The next day when they were at the church, they listened to a Congolese group singing "God Spell." It was so moving that the Sisters and even the Priest began to clap their hands and dance.

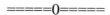

J.L. asked for the names of her mother's sisters while recording. She only knew the name of one of them, an aunt named Viola. The name *"Harriet"* was recorded. Later she found some old family records at the Courthouse. There was a photograph of a grave stone with "Harriet" on it. Harriet had died in 1893 at the age of four and J.L.'s mother had not known Harriet either, as she was born eight years later.

Later on the same recording, J.L. taped, *"Aunt Harriet loves Judy"* followed with the sound of a barking dog. Judy was J.L's pet dog who had died many years ago.

Researcher, Clara Laughlin, has made many recordings that have pertained to events that were unknown to her at the time. She was recording one day when she received, *"You should learn shortly I died and am in heaven. Surprise I went so early. You should tell Libby."* Clara had been a good friend with Libby and her husband when they had lived in another state. A day later, Clara received a card from Libby saying that her husband had died. The recording had taken place seventy-two hours after his death. Clara contacted Libby as requested and Libby found the recording very comforting.

In 1986, Clara received a message in her deceased husband's voice that said, *"Hello, grandmother."* Her daughter later found out that she was pregnant and the doctor told her she would have twins. Clara went back to her tape recorder and asked if this was true. Her husband Tom replied, *"Single baby boy,"* which proved to be correct.

Clara's daughter called to say that the father of a close friend was in a hospice and that death was imminent. She said that the family was distraught. Clara was asked for a prognosis of the situation. She told us that she printed the man's name on a piece of paper, held it to the mirror and asked her guide Dr. Callie for help. He replied, *"How about survival!"* A couple of days later the man was moved from hospice and was recovering.

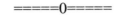

This last example of a precognitive EVP will give you a chuckle. Ann Longmore-Etheridge sent a box of books to her sister. Before her sister received the books, Ann was making a recording and asked if anyone had a message for her sister. She recorded a female voice saying, *"Yes, I have. Put your clothes on."*

Ann and her sister discussed the message and could not quite understand its meaning. Just a day later, Ann's sister was lounging around her living room in the summer heat with little clothing on. The doorbell rang and she peeked out to see who it was. It was the delivery man with the books. She was unable to get her clothes on before he left and so she did not receive the books that day!

Phone Calls from the Dead

A transcommunication group founded in Frankfurt, Germany, in October 1987, received excellent results. Their contacts were through tape recorders, telephones, television and computers. They reported on a woman's voice, which spoke over the telephone. The voice sounded like Hanna Buschbeck, the founder of the German VTF, who had crossed to the other side years before. She said, *"You will have eternal life after you pass over. You will be given the opportunity to learn, to see and your being will get closer to the truth. Do not be afraid of dying, for there is no death. Do not make the mistake of considering your reality as the only real one. It will make your later learning process more difficult."*

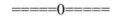

Susan Cole wrote that her friend Rick had died while she was away from home and unreachable. She realized later, that from the minute that he died, she had begun experiencing poltergeist activity. In retrospect, she came to feel that Rick had been trying to reach her.

She was packing for the trip to his funeral when the phone rang. She said, "Hello" and a voice said, *"Good Morning, how are you?"* She told us that the voice sounded strange. It had a lot of reverberation in it and sounded hollow, as in a tunnel.

"Fine, who's this?" Susan said. *"This is Rick, who's this?"* the man said. "This is Susan," was her confused answer. At this point she said that her mind was searching for whom she knew named Rick. She wrote, "I didn't yet even imagine it was my deceased friend."

"Hi Susan, how are you?" the man on the phone said, sounding very happy, lively and friendly. At this point Susan recognized his voice, but was very startled and thought that the call was some sort of cruel joke or strange coincidence. "What number were you dialing?" she asked. He said, *"5032" or "5033."* Susan told us that she was not certain which the man said, but that 5033 is her correct number. She replied to the man, "I think you have the wrong number." She told us that she knew no other "Rick," but her friend Rick who had just transitioned. The man then made strange gargling sounds and suddenly there was a dial tone. She dialed the 5032 number, to see if a Susan was there but it was not a working number.

"I then called a friend," Susan wrote. "The first thing that she asked me was if he had sounded hollow, like an echo. Boy, did that startle me! My friend told me that her grandmother had called after her death and that was how she sounded. Looking back, I realized that when I got upset during the call, my emotions broke the connection."

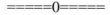

J.R. described an experience he had with a phone call, "A distant cousin was married to my father-in-law. I was aware of her being in the hospital and in poor health due to her age. One morning the phone rang, I answered and a voice said, *'She's Dead.'"* Thinking that this was a prank, John hung up the phone. Five minutes later the phone rang again. It was John's brother-in-law telling him that the woman had passed.

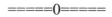

Pat J. ran a mediumship circle with her husband in the mid-1980s.[54(V15N4)] The day that Pat's mother made her transition, she received a telephone call from the late Helen Duncan. As many of you know, Helen Duncan was a famous physical medium during her life on earth. She was also the principle communicator from the other side in Pat's circle. Over the phone, Helen told Pat, *"Mum is safe and with us."* This was then followed by the voice of her son, who was also on the other side.

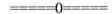

B.E. had made a phone call to a friend. The friend asked her to hold on a minute while she went to get something. All of the sudden, a man's voice came on the line and said, *"Hey, how are you doing?"* and then was gone. The voice sounded just like that of her deceased brother. She asked her friend if her son could have picked up the phone and was assured that he was away and that no one in the house had picked up the phone.

Jean's son, John, was a friend of a boy named Sam. Sam had been killed when he rode his bike into the side of a bus. A few days after Sam's transition, the phone rang and Jean picked it up. She recognized the male voice as Sam. He asked, *"Is John there?"* Not believing what she was hearing since she knew he was dead, she asked who was calling. He replied, *"It's Sam."* Jean asked for a phone number where her son could call him back and he answered, *"No, I'll call back."* There was no sound of the caller hanging up and the line just remained open.

=====0=====

Here in the United States, probably the most sensational and reported series of phone calls came from Konstantin Raudive. These took place in January of 1994. He called several people who had been instrumental in advancing the knowledge of ITC. Sarah Estep, Mark Macy, Hans Heckman, George Meek, and Walter and Mary Jo Uphoff were all blessed with a phone call from this eminent researcher who was on the other side at the time.

Sarah was working in her office when the phone rang. When the voice said it was Konstantin Raudive, she had the presence of mind to turn a tape recorder on that sits next to the telephone. She was a little shocked and said, "How are you Dr. Raudive?" His reply shows that his sense of humor was still intact, *"I'm as fine as a 'Dead One' can be. Dear Sarah, thank you very much for everything you did for the propagation of the voices. We tried and we succeeded in building this bridge to the States. You are one of the first who are contacted by this means. Thank you very much for all the work you did. We are very proud and honored that we could contact you. I must interrupt now. This was the first contact, this is Konstantin Raudive."*

Mary Jo Uphoff told us about the phone call that they received, "Raudive called one morning and Walt answered the phone. Walt motioned to me to take the extension telephone in the adjoining den. I was in a rather obtuse state of mind and at first didn't know what he wanted and by the time I picked up the den phone the conversation had ended. I came back to the kitchen where Walt had hung up and asked, 'Who was it?' 'Raudive', he said. He told me that Raudive had said that he wanted to get a call through and that they were experimenting. It took me several moments to realize that this was a most extraordinary event. One of the characteristics, in my experience, of such events is that when they are taking place, there is not any sense of awe or strangeness; it is only afterward that it seems to sink in that they were very, very extraordinary."

EVP on Answering Machines

There are many reports of EVP voices appearing on answering machines. The first such message we received was rather amusing. Our family room is open to the kitchen with a long counter in between. The phone sits on this counter. The family room did not have a chance of being used for its intended purpose. Instead, it is filled with two desks, file cabinets, and other legacy of the AA-EVP.

The first answering machine we had used a micro cassette tape and we were sure that it was just exactly the right device to capture an answering machine EVP. But we had received no EVP on the device for six months and the telephone part was old and people had difficulty hearing us. Giving up the hope of receiving EVP in this manner, we got rid of it.

In less than a week after purchasing a portable phone with a digital answering device, we came home to our first phenomenal message. The message began with a long period of silence and then a loud whispery voice saying, *"You need an office."* Most EVP have a different sound to them as compared to a human voice. Once you have been recording EVP for a while you immediately recognize them and what was on the machine was definitely EVP.

From the following examples you might believe that it is a good idea to check your answering machine for EVP. If your machine is like ours, there is no tape, but you can buy an inductive coupler from Radio Shack (Part Number: 44-533). It looks like a suction cup with an audio cord attached. You place this over the speaker of the machine and record messages from your answering machine into a regular recorder or directly into your computer. You will also learn that if you can download and listen to the messages left by your friends you can occasionally find EVP in the pauses between their words. Our friends now know that, when they reach our answering machine, they should leave a message and then leave a little time at the end before they hang up so that the entities can also leave a message.

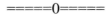

L.S. had taken two jobs during the summer and she also attended classes. This busy schedule had prevented her from conducting her normal EVP recording sessions. One of her jobs was to take the messages off of her employer's answering machine. On the last day she was to work at this job, she received a strange message on this machine.

She had been out of the office for twenty minutes and came back to find one message on the machine. She recognized the voice on the machine as that of a person that she had recorded in her EVP experiments. It was unlike any normal message and the voice did not give a name or whom he was trying to reach. He spoke quickly in a monotone voice saying, *"We wanna make sure you come back and visit. We love you."*

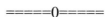

D.S. has received numerous messages on his answering machine. He says that they often appear before the person calling begins talking, in between breaks in the person's words and at the end of the recording. *"I'm missing now my pretty love,"* is one of the many messages that he has recorded on his machine.

D.S. receives calls that have a long silence in which the entities then leave messages. This is exactly what happened with our first answering machine EVP. D.S. once said that he felt that these might be telemarketer calls, or automatic dialers used for the same purpose, that initiate the call and then do not immediately hang up. This type of call provides a way for the EVP messages to come through. It's funny to think that there could be anything positive that could come from a telemarketing call, but it sounds like this is the case, as long as they are intercepted by your recorder.

One more recent answering machine EVP that D.S. has received was of a mature sounding female who said, *"We walk alone, please pack."* He surmised that, "Perhaps this means that each person must be fully prepared in his or her journey through life. Maybe it is a reference to the Spiritualist concept of personal responsibility."

L.C. wrote, "My mother had a phone call from the dead about ten years ago. She was checking her messages on her micro cassette answering machine when she received a message saying, *'Tell Laurita ... tell Laurita, tell Laurita.'* It was my Aunt Stella's voice in monotone, but that was my aunt's voice. I heard it for myself. We thought, 'What sicko could do this to us?' We all missed my aunt. My mom was very close to her. Why would someone fake my Aunt Stella's voice and say my name but not leave a clear message? At that time I was going through a lot of struggles in my life and it was a very hard time for me. Now I know about EVP and such things and I know that Aunt Stella was simply telling my mom to tell me everything would work out."

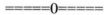

Bill Weber related two messages that he received on his answering machine some time ago. In the first, a male voice recorded, *"We've been excused."* This voice came on the machine with only one phone ring; the machine was set to pick up only after four rings. On another occasion Bill's machine recorded a female voice singing, *"Bill."*

When Dave Sircom was conducting a regular recording session, he recorded, *"We tried to call you."* He immediately went to his answering machine with a microphone in hand. His machine had recorded two messages, both of which were hang-ups. By listening to them with audio editing software in his computer, he was able to hear the messages, *"I saw the whole house"* on the first and his name *"David,"* on the second.

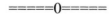

Perhaps Sarah Estep should get the credit for receiving one of the most memorable EVP message on an answering device. The message was received in 1995 and came from Konstantin Raudive. This was recorded on the telephone company's answering service and not on a personal home answering machine. The call was not only very evidential in proving that the other side is aware of what is going on in our lives, but was also a great source of comfort to Sarah at a time that she needed comforting.

Sarah's husband, Charles, had suffered several strokes. She had taken him to the hospital for tests the day before the call from the other side came through. The information was not what they had hoped and Sarah was disheartened and a bit down. The next day, she returned from being out of the house for a couple of hours and checked the answering service. The message that she picked up from the telephone service was a voice that she had heard before. It said, *"Dear Sarah, this is Konstantin Raudive. Thanks a lot for the engagement in ITC and its recent development. Our medical group here on this side is working on intertwaining* (intertwining) *lots of life and afterlife of Charles. I can assure you that every possible intervention will be made. This is Konstantin Raudive."*

Unexpected and Expected EVP on Some Unexpected Devices

Most researchers working in EVP now know the name Charlotte (Charli) Claypool. When Charlotte Claypool first wrote to us about EVP and about her interest in joining the AA-EVP, we were immediately interested in hearing more about her EVP contacts. You see, Charli has been receiving EVP messages in a number of ways but most interesting are the voices coming from her coffeepot!

Charli wrote, "I am a Real Estate Broker with no prior interest in EVP or paranormal activities. Phenomena began taking place in my presence ten months ago. Voices began to emanate from a common household appliance which was purchased in November 2000. The voices are discernable and can be heard in 'real time' by anyone; they do not have to be recorded to be heard. The entities coming through are prolific speakers and they are numerous. The "common household appliance" is a Krups Duofilter, Chrystal Arome Coffeemaker.

"Once I got up the courage to record the voices, using a digital voice recorder, I found that the entities actually had discernable male or female voices, different dialects and plenty of opinions. They readily answer questions. The digital recorder downloads directly into the computer for analysis.

"I had these ladies that kept talking about pink lemonade so I went out and got a bottle and set it by the coffeepot. Later I asked if they knew about the pink lemonade. The Ladies in unison replied, *'Yes, sweet new girl.'*"

Charli said she gets different groups through different media. "Direct recording off my ham radio with scratchy static produced relatives; people who knew my mother." One interesting set of EVP was in regard to Charli's grandfather, Whipkey, whose nickname was "Art." The voices have spoken about her grandfather using both his last name and his nickname. Charli never knew her grandfather and had no idea what name he went by. He was estranged from her grandmother decades before she was born and her grandmother refused to ever speak about him. They have said, *"That's Charlotte Whipkey's young lady."* Charlotte Whipkey is her mother's maiden name. On another occasion she recorded, *"I see Charli's Mother,"* and then another spirit voice said, *"No, that's not Charli's Mother, that's her Grandmother Gladys."*

Charli's husband is adopted and her spirit friends have referred to him by using his first and last birth names, which are different from the name given him by his adopted family. She also writes her questions down on a piece of paper for those on the other side to read. She wrote, "They love the notes. One lady said, *'Fun to read letters from Charli.'* I knew that they could read because of different remarks referencing reading."

=====0=====

There are several other AA-EVP members who write messages and notes to their spirit friends. Clara Laughlin often puts her notes up to the mirror she uses in her recording room. Her friends have told her that they "see" through the mirror.

Also, we want you to know that the voices have come through with many interesting background sounds. After we ran a story in the NewsJournal on Charli's coffeepot, many members wrote to tell us of other background sounds that produced the voices. Some of them were a child's swing, a steam iron, and the sound of material rubbing against other material. Many members have used the noise of an electric fan or running water as background noise to help the voices come through. One member always records when she fills the Jacuzzi.

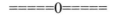

Ann Longmore-Etheridge related an interesting encounter with EVP. "There is a small parking lot by the cemetery where I usually park my car. I always have a feeling that someone is watching me from the cemetery whenever I park and walk across the road to my house. I've always thought of this 'watcher' as an angry man."

Ann's son has a two-inch long battery operated memo recorder that is part of a toy key-chain. She had just brought her son home and was playing with the recorder as they crossed the street from the parking lot. "First I recorded myself saying, 'Hello,' and played that back. Then I said to my son, 'I think I'll take this to the cemetery and see if the ghost will talk to me.' Then I hit the record button accidentally, actually meaning to hit the play button to hear myself saying, 'Hello,' again. When I did hit play, I heard a deep male voice say, *'No!'* I guess there really is a grumpy man in the cemetery who doesn't want me around."

Psychokinetic Effects

Although it is not often reported, some unusual things can happen during EVP recordings. Both of us have experienced hearing direct voices and we have had our equipment turn on by itself. On one occasion, we had just spent a week of vacation in an intensive psychic development class. When we returned, we sat some equipment up on the floor and conducted an EVP session. Lisa was sitting on the floor between the tape recorder and the microphone. She leaned over the top of the recorder to look at the counter and received quite a surprise. She yelled and quickly moved her head back. What felt like an ungrounded 220 volt circuit had buzzed right through her head.

=====0=====

Bill Weber has recorded EVP using conventional methods for many years. A few years ago, he began experimenting with a piece of software call EVPMaker.[72] He was quite excited at first, as he began having recurring contacts with a group of what seemed to be teenagers or young people. They used words like "dude," gave their names and commented on many things. The contacts soon turned critical about Bill and things like his housekeeping skills. The EVP messages contained more and more profanity. Then things began occurring in his home. Lights would go on and off, the doorbell would ring when no one was at the door and his sleep became disturbed.

It is known that prolonged, careful analysis of audio tracks, while listening for low-volume EVP, can have the tendency to enhance a person's clairaudient and clairvoyant perception. So, we were not surprised when Bill stopped recording for a while. Despite Bill's strong background in spiritual matters, the group of entities he was apparently in contact with via EVP was causing too many disturbances in his life.

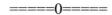

M.H. was experimenting with using his ultrasonic burglar alarm as a noise source during recording. He had been recording for only a couple of minutes when the front door, or storm door, was violently shaken. Irritated that the person making the noise at the front door had wrecked his recording, he turned the recorder off and went to see who it was. No one was there.

M.H. resumed taping and five minutes later, the same thing happened. Being prepared this time, it took him only seconds to reach the door and throw it open. Again, no one was there.

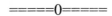

C.G. conducted her recording experiments in a book store that she operated. The recordings were done near a window in the front of the store. Hanging from the window were many crystal prisms. One person often spoke on her tapes and always said, *"I love you."* Whenever C.G. was listening back to this particular voice the crystals would start clinking and banging the window. She wrote, "I know he is doing this. It is a physical manifestation of his presence."

EVP Used for Investigations

EVP has been used to find missing items and has even been used in investigations of missing persons and to help solve crimes. Perhaps in the future it can be used more for this purpose. The use of EVP to help solve crimes poses some interesting problems which are similar to those posed by the use of a psychic or medium. Some psychics are better at this than are others and many have done outstanding work that has provided a great service to government agencies and families. In the same way, some EVP experimenters are more capable than are others and their messages are more reliable.

In the use of mediums, the problem comes with the inexperienced mediums who contact the media or authorities to report predictive messages that rarely come true. This problem is aggravated when it comes to receiving this type of information through EVP. In EVP, the message can be spoken in a very convincing manner by an entity who claims to be the victim or someone who should know about the situation. Yet, unless the experimenter has established a working, trusting relationship with the entity, either directly or via a "control" or "gatekeeper" entity, the message may be initiated by a mischievous individual on the other side who might be interested in nothing more than being entertained.

Just because a person exists in another dimension, that person does not necessarily have the ability to know more than we do here. It is certain that an EVP experimenter could have contacts on the other side that are as good at solving crimes as do the few excellent mediums and psychics that we have on earth who are doing this work. But without a trusted gatekeeper, there is a fair chance that EVP messages concerning something that the experimenter wants to know may be counterfeited in some way by a mischievous communicator. Unless a bridge to a group of known entities has been established, the communicating entities are often strangers to the experimenter. Would you trust a message from a stranger you might meet in the supermarket? Would you deliver such a message to the police or to a grieving family?

Prediction after prediction has been brought to us from various people who have not taken the time to understand with whom they might be communicating. It is natural for an experimenter to want to share their message with others. Sharing messages in a forum, such as the AA-EVP Idea Exchange or amongst friends is part of the learning experience the Association offers to members. However, it is quite another matter when the experimenter with the predictive message decides to call the media. Sadly, we have known people who have told us about their predictive EVP, and even though the predictions had not come to pass months or years later, we have seen them speaking to the public about a new predictive EVP. The media loves sensation and what do they care if only an occasional prediction comes true.

The lack of discretion in an experimenter can cast a shadow over the work of other EVP and ITC experimenters who are patiently building a foundation of evidence and credible examples. These predictions that never come true are used by skeptics to point out that EVP researchers are hysterical and unbalanced individuals who are hearing voices that are not there.

If you are interested in EVP and if you think it could be used to help solve police cases, we ask that you first have a strong connection with your EVP partners on the other side and that they prove their ability in doing this kind of work to you before you go to the media. Remember, you are representing all of us when you speak about EVP to others. One excellent source for raw material necessary for developing this ability is the "Secret Witness" program that most communities have. It is usually possible to gain access to public information about many cases, which can be used for "targets" in EVP experiments. Then all that is required is the patience to track the outcome of these cases and to compare these outcomes with your results.

With that said, here are a few examples of where EVP did provide some interesting information.

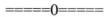

An article was published in the January 1987 *Gazette* titled, "Spirit Voice on Tape Recorder Traps Murderer." The *Gazette* was published by the Spiritualist Association of Great Britain. Two pensioners moved from Switzerland to live in Vienna. Their new neighbors liked the couple and they made many friends. After not seeing the couple for four weeks, the neighbors finally called the police. The bodies of the couple were found in the house and both had been strangled.

One of the neighbors had been successful for many years in recording voices. He gathered a group of friends and they sent their thoughts out to the murdered couple. The tape recorder was turned on and they asked repeatedly, "Do you know your killer?" A voice finally came through with the answer, *"Yes."* "Do you know his name?" they then asked. The same female voice answered, *"Bozidar Sajn."*

The friends were convinced that this was the voice of the wife. They took the recording to the police who laughed at them. However, the police later came across a man with the same name who had also lived in the same block where the crime took place. He confessed and received a life sentence.

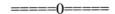

Bill Weber made a recording two days after John Kennedy's plane was reported missing. The Navy and Coast guard were searching extensively trying to locate the plane, and Bill thought that he might be able to discover the location of the plane through EVP. He repeatedly asked for those on the other side to tell him where the plane was, but none of his questions were answered. However, at the beginning of the recording, when he first asked for the location, a voice replied, *"He's here."* Bill asks John Jr., "How are you sir? Do you have a message for the world?" Immediately a loud clear voice answered, *"I live in Spirit!"* Bill continued to ask for the location of the plane and a voice replies, *"John was killed."* A minute later an even louder and clearer voice repeats the message, *"John was killed."*

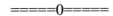

Clara Laughlin had a friend who was a psychic. Her friend was working with the police department on a case in which he had been a friend of the murdered victim. Using her tape recorder, Clara recorded an EVP with the middle name of the murder victim and this later proved to be correct. She also recorded a woman's voice saying, *"Ralph, I miss Ralph."* It turned out that this was the name of the murder victim's boyfriend.

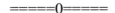

Erland Babcock and his son, David, were asked by the aunt of a girl who had been murdered in her home to "look into" the crime scene. Erland took his recorder and recorded two of the better EVP examples we have in our files. When Erland and his son played back their recording, they were startled by a loud woman's voice saying, *"This is my dream."* The EVP voice was actually louder on the recording than the voices of the investigators. The EVP had absolutely no meaning to Erland and his son but they decided to play it for the Aunt anyway. When the Aunt heard the voice she became very excited and rushed into another room. She had a piece of paper in her hand when she returned. The girl had written a poem right before her death titled, "This is my dream." The Aunt also recognized the girl's voice.

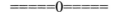

We have heard from many individuals about EVP helping in small ways. This is one such story.

Jürgen Nett, the chairperson of the German ITC group, VTF, wrote an interesting article titled "The Rediscovered Safe Key," for the VTF quarterly journal. Thanks to George Wynne for translating the story for us.

"During my talk at the Kolpinghaus [During a VTF Annual Meeting], I reported a happening that was confirmed by Mrs. Werner of Rotenburg, Fulda, in the presence of all the participants. In the meantime, Mrs. Werner has passed into the spirit sphere.

"What happened was the following: Mrs. Werner came to my office one day and requested that I make a recording for a well-known businesswoman in the town of Rotenburg. Before leaving for her vacation, this lady had hidden the key to her bank safe deposit box. She always used the same spot for this purpose. On her return from the vacation she looked for the key and could not find it in the usual place. She searched the entire house. The key simply could not be found. She was afraid to tell her husband, since a replacement key would cost a considerable amount.

"In her plight, she contacted Mrs. Werner because the latter was always engaged, "in such queer doing as talking with the dead and similar 'humbug,' as this is called in the small town. I made a recording for Mrs. Werner and the answer came: *'On top of the wooden shelf.'*

"With this bit of information, Mrs. Werner visited the business woman and shared the sentence with her. The woman thought about this and replied that she did not have any wooden shelf, except in the bedroom above the two beds where linens, towels and washcloths were put away. But, she had never put the safe key there. Nevertheless, they went into the bedroom, and what do you know? Under a pile of towels was the missing safe key!"

EVP and Reincarnation

From the examples of EVP that you have read, you can see that for the most part, those from the other side who record their voices onto our recorders are those who have lived on earth and are considered by us to be amongst the so-called dead. A couple of researchers, one being Sarah Estep, have been told by communicating entities that the entity and the researcher knew each other in a past life.

After Sarah began having good results with recording the voices, she returned to her exploration of reincarnation, since that was how she had first tried to prove to herself that we survived death. During this time, a clear male voice came through and said his name was Jeffrey and that he was her brother in a previous life.

Jeffrey told Sarah that they had been brother and sister in the 1700s in Philadelphia, and that he had worked as a lamp lighter. In her research, Sarah found that Philadelphia did have men whose job it was to light all the lamps in the city when night came. In addition, Sarah's grandfather, James Wilson, had come from Scotland during that time. He was a tutor at the University of Pennsylvania in Philadelphia and also an attorney. Her grandfather, along with Benjamin Franklin, began the Law School at the University. He had a wife and several children and lived most of his life in Philadelphia.

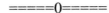

One of Carol Barron-Karajohn's first messages was *"I love you Kathy."* She thought that it was someone talking to someone else. Experimenters often record voices that sound like the entities are talking to each other and are not even aware that they are being recorded. This is much as you would experience if you picked up your telephone hand set on an old time two-party line and heard two people carrying on a conversation. However, Carol was soon to learn that they were indeed addressing her.

Not long after Carol had begun experimenting with EVP, she noticed that some of those who left messages on her recorder referred to her as Kathy (Cathy). She also discovered that she had a primary spirit communicator who gave his name as Hegeler. Hegeler told her that he was William Hegeler and that he was German. Later, Carol met the German EVP researcher, Ernst Senkowski, at a conference and asked him about Hegeler. Ernst sent her a genealogy report on the author, Wilhelm A. Hegeler. Carol also learned that Hegeler had a sister named Catharina, which was interesting because Carol had been called Katiana on her recordings, which she thought might be a pet name for Catharina. The entities have told Carol that Hegeler is her brother and he has told her that she is his sister. When she asked if she was Kathy in a past life, they replied, *"You ARE her!"*

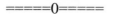

More recently Charli Claypool wrote to say that she really was not into past lives, as she had her hands full with EVP and its overwhelming proof of spirit existence, but that her contacts on the other side had told her that, *"Your name is really Valerie Davis. You were a really old slave."* Charli was very surprised at this declaration as she has ancestors on her father's side with the last name of Davis.

EVP Coming from Other Dimensions

Many EVP/ITC researchers have had contact with entities who say they are extraterrestrial beings. Voices are recorded and Video ITC images of beings that do not resemble humans are numerous. Some of the voices that we have recorded say things that make us feel that they may not be of human origin. For instance, a couple of the messages of this type that we have received are, *"We come in a Spaceship"* and *"We are from Alpha Centauri."* One time, in the days of our early recording, we asked where they were speaking from and recorded, *"From a spaceship."*

Over the years, we have recorded voices that sound as if they are computer generated. In fact one even said, *"Now you have the computers."*

Many will probably recall that Sarah Estep recorded many of these alien voices. In fact she has said that ten percent of her voices were of extraterrestrial origin, and that they have mentioned Alpha Centauri more than once. There were two chapters devoted to these contacts in her book, *Voices of Eternity.*[3] In that book, Sarah said that they were usually of a better and louder quality as compared to the spirit voices, that they often used words not found in the dictionary and that some voices had a computer generated quality.

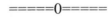

It is interesting to note that Dan McKee, a person whom Sarah felt was one of the best tapers in the United States before his transition, also recorded thousands of space messages. One of Dan's recordings said, *"Tomorrow night you will be looking for Centauri."*

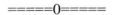

Jutta Liebmann, of the German VFT, has also taped many loud and clear voices which she feels are non-human. Some of her contacts have said that they are from the Alpha Centauri region and others have said that they are from the Sirius system.

Sarah, Dan and Jutta did not know at the time they received these messages, that other researchers had received similar messages. Although we did not recall knowing about these contacts at the time we recorded the Alpha Centauri message, we must have known about Sarah's messages, as we had read her book.

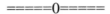

In 1993, Sarah Estep received an interesting letter from Jamal Hussein, a physicist working in the Paramann Programme Laboratories in Amman, Jordan. He wrote at length about contacts that he and his colleagues had with inhabitants of other dimensions, "After over a year of continuous communication with an unseen voice, we were astonished to hear this very voice, which we used to think of as being a human voice of someone who lived a life on the Earth Plane, announce that he was very ashamed that he had 'cheated' us by claiming that he was a human being because, in fact, he is one of the inhabitants of an unseen world that surrounds our world. He and his people can take human form when they like and the reason he claimed that he was human in his contacts with us was his fear that we might end communication with him on the basis of being afraid of him, as he does not belong to our plane of existence! Further communication has resulted in proving that what he was saying about his real identity is absolutely correct. I feel that these beings are very emotional and that they do, as you have mentioned in your book, express love and friendship and a desire to help. We in the Paramann Programme Laboratories, are very happy to know that you are brave enough not to be afraid to announce what you have come across in your over sixteen years of tape recording unseen voices. I think that you, unlike those who try to interpret all paranormal phenomena as being spiritually caused, have the courage to declare that you have experimental evidence that can prove the existence of unseen worlds, which have nothing to do with human spirits and whose inhabitants are endowed with consciousness which is by no means human though it does have a lot of similarities with human consciousness."

Some Cautions about Working with EVP

In the Summer 1988 AA-EVP NewsJournal, Sarah ran a brief review of the CETL[62] newsletter written by Maggy and Jules Harsch-Fischbach in Luxembourg. The couple expressed that experimenters could attract positive as well as negative contacts. They quoted a comment made by Konstantin Raudive, "Transcommunication is not a hobby for people who can't cope with the realities of life."

In the heyday of Spiritualism, people sat in groups trying to communicate with the other side. One of the important benefits of these groups was to provide a safe place for mediums to develop. For instance, when a person is working to develop trance mediumship, he or she can feel open and safe if there are experienced mediums in the circle who are able to offer assistance should there be trouble. Many Spiritualist churches offer development groups today.

There have been a number of occasions in which the value of a circle for developing mediumship ability has been demonstrated. For instance, we worked for several years with the Reverends Sandy and Gene Pfortmiller and their group from the Church of the Living Spirit in Phoenix, Arizona. During one trance class, a relatively inexperienced medium was "taken over" by a lost little girl. In this instance, the student medium behaved as if she had difficulty moving her attention away from contact with the little girl. It was not as if the little girl had taken control of her body. It required quite a bit of effort on the part of the Pfortmillers to first assist the little girl, and then to bring the medium's attention back to her body.

In another instance, and with another development class, a more experienced trance medium was "taken over" by an entity who yelled that, "I (the entity) am now in control!" It was actually quite shocking but the facilitator of the group remained very calm and maintained control protecting the medium and sending the entity on its way.

Many, if not most, EVP experimenters experiment with EVP alone and do not have the benefit of a group or another more experienced EVP researcher. It would be best if everyone who experimented with EVP had a strong metaphysical or spiritual background. One recommendation we would make is that EVP experimenters continue to learn everything that they can about what it is like on the other side through reading and that they also try to attend spiritual development classes.

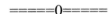

In the Autumn 1989 NewsJournal, Sarah Estep wrote, "Obviously, no one starts taping, automatic writing or playing with the Ouija Board, thinking he or she will become possessed or obsessed. But it can happen. As I wrote on page 196 of *Voices of Eternity*, The difficulty is in ascertaining who is susceptible.... I am not trying to frighten people who are thinking about beginning to tape.... Working in the field can bring some of the most rewarding, enriching experiences of your life. It would be amiss of me, however, not to caution you about the darker side of psi. Experienced tapers will tell you not to believe everything that comes through. You have your liars on the other side, as you have here. The more they realize they are getting you upset, the more they will continue. Human nature being what it is, we may imagine certain messages are there when in fact, there is nothing. The important thing though is, if a person believes a message is on tape, he or she will respond to it for better or worse.

"If anyone at anytime thinks he or she is in contact with low level entities, leave your tape recorder. Remain in control at all times. Anyone can sit down to tape. It takes much more inner strength to pull the plug and walk away."

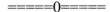

Jutta Liebmann, of the VTF, once wrote that it is necessary to caution new tapers, saying that some might think that all those who are on the other side have become angels in the Spirit World, and that this is really not the case. She wrote, "They remain the same personalities as they had been on earth."

=====0=====

Back to thoughts on spiritual development, Spiritualist training teaches that as you work toward the development of your abilities, you will come in contact with a guide or what some call a door or gatekeeper. This person protects you and the contact field. The "contact field" is the etheric energy which helps to power nonphysical to physical communication. The same thing can be said for EVP. As you work with EVP you will build up a contact field and if you are sincere and not just approaching EVP out of idle curiosity, you will attract to you beings of a higher awareness whom you will learn to love and trust and who will help protect the doorway that is open to other worlds.

Lisa remembers, "When we first began recording, Tom had also started writing a book. Sometimes, I would conduct EVP experiments on my own, giving Tom a chance to write. Early in these recording sessions, I found myself in a most disconcerting situation. What I would call a 'dirty old man' began showing up on the recordings. He clearly enjoyed the opportunity to speak on the recording equipment. He not only used profanity, but also was quite vulgar and frequently made sexually suggestive remarks. What was even more unnerving was his ability to tell me exactly what I was wearing and then say, *'I'm right behind you.'*"

The "dirty old man" that bothered Lisa via EVP messages was very obnoxious. He tended to hog recording sessions and was extremely persistent. You name the trick to make him go away and we tried it, from the 'go to the light' speech, trying to reason with him, trying to educate him, to finally demanding that he leave. Prayers, white light and positive affirmations failed. In the end, everything that we tried just seemed to give this uneducated spirit more power. Nevertheless, EVP was so phenomenal and held so much promise that we were determined to regain control over our new spirit phone.

What did we do, you ask? Well, we gave up and ignored him. And instead, we concentrated on the positive messages from a wonderful group of spirit ladies who continued their effort to place their voices on our audio tape, despite the interruptions by the rude man. The "Ladies" helped by warning us that he was about to speak, and the second that we heard his voice, we simply ignored it and moved on to the next EVP message. Of course, we did listen to some of his messages and we knew that he was complaining about our treatment of him. Like some sort of bizarre phantom, we could actually hear his voice growing weaker until finally he melted away into the ether like the bad witch in the Wizard of OZ. There have been no similar problems with unwanted communication after this episode. With the help of our spirit friends, we had finally built a secure bridge of communication that has held to this day.

Letters have come to us from people who belong to ghost clubs or paranormal groups telling us that they would never try recording EVP in their own home. They indicated that they feared finding messages from one of the angry or mean entities that they come across while recording in haunted sites. First of all, ninety-nine percent of the entities that we have recorded in haunted sites have been nice spirit folk, perhaps confused, but rarely scary as portrayed in the horror movies. Secondly, we have never once had an entity from an investigation turn up in a recording made in our ITC experiment room. Nevertheless, if you have any fear or doubt about experimenting at home, then do not. Our thoughts are powerful things and will attract to us our beliefs.

=====0=====

Charli Claypool wrote to provide some thoughtful information about how to deal with negative entities, and to offer another way of looking at this type of situation. The letter was in response to a discussion within the AA-EVP email sharing group. The group was discussing an angry entity that had been coming through on a particular member's recordings. Charli wrote, "Spirit entities with this kind of personality change radically when treated with kindness.... It takes some intentional fortitude in the beginning to be kind. I've heard the disappointment in the voices of entities who have had their feelings hurt by 'banishment.'

"What bugs me is that spirits for the most part have tried to interact with their families left behind on earth and have found themselves shut out by their loved one's inability to see and hear them. We, who can penetrate the veils and interact with these personalities, have the opportunity to show them love and improve their lot. Remember that love is the key driving force to interaction between the veils."

Charli deserves credit for maintaining such a positive point of view and we agree with her thoughts. There is a great need for rescue work, which involves the efforts of the "living" to guide those on the other side who might be "stuck" so that they may move away from our locale and to their destined worlds. Just like hospice work, there are not enough people drawn to this calling. There may be some who are not as strong as Charli and who are unable to give the kind of kindness certain entities need, but all of us have a responsibility to help when we can.

Not all entities you will encounter want to be rescued. There is substantial evidence that some entities are, indeed, earthbound, and that they are so because they do not understand that their physical body has died, because of unresolved issues or because they feel that they will be punished for something they did in the physical if they go on. These entities can often be reasoned with and convinced that it is both okay and important for them to go on, and reportedly the most effective way to communicate what "going on" means is to tell them to go to the light. However, there is also evidence that some entities chose to remain close to the physical. They have free choice and we can only offer our advice.

Most people tend to pick their friends on this side of the veil very carefully. As far as our close friends go, we pick those with a positive attitude, similar interests and ethics. Finding people with whom we can learn new and interesting things, and with whom we can share what we know, is indeed special. This seems to create growth and bring more positive things into our lives—and theirs. This is what we seek from our friends on the other side as well. Our aim has been to reach for the highest information possible.

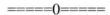

Before his transition, Bill Weisensale published a technical EVP newsletter for many years called, *Spirit Voices.*[73] He wrote, "Electronic transcommunication is awesome and fascinating. In time, without realizing it, one can become obsessed with this work. If you find taping is beginning to have a detrimental effect on other aspects of your life, slow down or even stop altogether.

"If you hear things you don't want to hear, remember the 'off' switch is on our side. Like obscene phone calls, continuing to listen only encourages the caller.

"Always use common sense and take what you hear with a grain of salt. If someday, someone who appears to be your aunt tells you to sell all your property and invest in junk bonds, or do yourself in because life is more pleasant over there, forget it!

"Never under any circumstances consider taping as 'entertainment.' To do so, invites lower level Spirits into your home who may entertain you in ways you are not ready for."

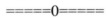

In an article written for *Psychic Observer & Chimes,*[74(460-466)] Harold Sherman, recounted a friend's distressing experience in trying to reach his wife and running into what he called "low grade discarnates." When the man lost his temper and told them to go away, it only seemed to increase their interference. Sherman listened to some of the recordings and called it pure drivel.

Sherman came to the conclusion that the mental attitude of the experimenter played an important part in what was recorded. He wrote, "It is as though low grade intelligences come in on the emotionally disturbed 'wave length' of the operator. Unless the mind is prepared through prayer or a spiritually motivated meditative period, one is apt to invite the wrong kind of communicants."

Chapter 6

More Examples of EVP

Neither of us had heard of Electronic Voice Phenomena before reading Sarah Estep's book, *Voices of Eternity*[3] in 1989. It seems bizarre that we would believe that talking to the dead was possible. When we look back on it today, the idea of recording dead people's voices on tape ... well, it just sounds like no one should have believed such a claim. But, we did.

When we decided to try to collect EVP voices for ourselves, we began by following the instructions for how to go about it that Sarah had written in her book. Lisa lead the way by deciding that we would record at the same time each day at 7 p.m., and experiment everyday for at least a week. There were no voices on the recordings the first two days. The third morning was a Saturday, and Lisa woke up with the idea and compulsion to gather crystals that we had found during our rock hunting excursions, and place them around the recording equipment. She did not act on this right away, and went about her weekend chores; however, the idea of using crystals continued to nag her. Finally, she abandoned her chores and gathered some of the biggest crystals from the bookcases and shelves where they were displayed. She placed them around the equipment in preparation for our 7 p.m. experiment. When she played back the tape from that evening's experiment, she heard a faint voice say, *"Crystals help."* To quote Lisa, "I will never forget that moment! I played the faint voice over and over again. I had trouble sleeping that night and for several more. I lay there, eyes wide, staring at the ceiling, thinking about the significance of what I had heard and what that voice meant to my idea of reality."

=====0=====

In 1983, longtime EVP researcher, Carol Barron-Karajohn, read an article over lunch in the *National Enquirer*. The story was about a lady named Sarah Estep who routinely made contact with the dead via her tape recorder. Carol wrote, "Having had many paranormal experiences over the years, I was open to stories about the Spirit World. I could well accept visual visitation. But ... coming through over a tape recorder? This seemed a little strange, even to me!"

The story that Carol had read could have easily ended on her lunch break, but that was not to be. Carol had just begun a new health regime that included exercising with an exercise cassette tape as a guide. The tape recorder was right there from her morning work out and she knew that the other side of the tape was blank. She flipped the tape over without giving it much thought and pressed the record button.

After a few short minutes of recording, she played the tape back and was shocked to hear, amid the whir of the tape recorder, an extremely loud guttural whisper saying, *"Cah-Wall."* It was not exactly "Carol" but very close. She told herself that it had to be static and tried another recording. This time there was no mistake, she heard what sounded like a church bell, and then in a clear whisper, a voice that said, *"I love you! Do you love me?"*

Carol wrote, "Looking back, I should have been more hospitable, but I was overwhelmed and to tell the truth, a little scared to think I was being watched by the unseen! I nervously put it aside and decided to leave it alone. After a few days, I started to think that it was only my awareness that had changed. They had been aware of me all along, so what was there to be afraid of?

"This was the beginning of my relationship with those on the other side. To this day, I still feel overwhelmed at times, almost to tears, when I get a particularly significant message. But it is a feeling of joy ... not fear."

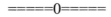

David Sircom first tried to record for EVP in August 2001, at 9:30 p.m. However, he recorded no discernable EVP on that first attempt. At 11:30 p.m., he again recorded, beginning by stating his name and asking, "Is there any friendly entity that wishes to speak?" David wrote that, "In playback I expected nothing. But, clear as a bell, I heard a voice say, *'Did anybody answer?'* Startled, I threw down the earphones and backed toward the other end of the room with all the hairs on the back of my neck standing up. After the initial shock, I tried again with my greeting and recorded a male voice saying, *'No time for this.'"*

David has gone on to record many other spirit voices. He meets with an EVP group and wrote to us saying, "If I never receive another EVP, I know they are there! It is because of your efforts and guidance that this one time skeptic is a full time believer!"

There Is No Death and There Are No Dead and They Want Us To Know This!

Marcello Bacci, a famous pioneer of EVP and ITC research, received the message, *"Life beyond death, beyond the life we know, from death is life."*

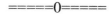

Clara Laughlin wrote, "The so-called dead want to communicate, they can communicate and they do communicate. They are not in the cemetery."

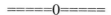

Carol Barron-Karajohn recorded an EVP that we never fail to play when we conduct a presentation. You would have to hear the enthusiasm in this voice to understand why we love it so.

Carol had made friends with another AA-EVP member. She wrote, "He called me many times because of my work in the field and we exchanged many audio cassette letters." A short time after her friend crossed to the other side, Carol was making a recording for EVP in the morning. On that recording she received the message that this friend would, *"Be on this evening."* Carol had never recorded an "appointment" for a friend from the other side to come through. That night she eagerly made her recording and was not disappointed with the results. The friend loudly and enthusiastically said, *"I'm alive!"* His voice was recognizable and just as it had been in his earth life. The man's relatives heard the recording and confirmed that it was his voice. He gave her other evidential information, such as where his parents had lived while on earth—information that Carol did not know. Carol said, "I'm sure that he is happy to be spreading the word from the other side that we don't die!"

A couple of other messages that Carol has recorded that seem to fit here quite clearly are, *"We, the dead, speak."* and *"Praise be the dead, the dead didn't die!"* Most researchers in EVP have recorded many messages along these same lines.

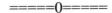

David had a friend, Bob, who did not believe in EVP at all. The two friends would argue for hours about the voice phenomena. While recording one day, David received the message, *"I'm across David ... Bob T ... Free!"* David thought that Bob was still alive but immediately began checking and learned that his old friend had died a short time before the message was recorded. The friend, who had not believed in EVP and had argued that it was not real, used EVP to prove his survival.

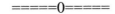

The famous German researcher, Hans-Otto Koenig, recorded the name of a close friend who was in the hospital. Fifteen minutes later, the man's wife called Koenig and said that her husband had died fifteen minutes earlier! After this, Koenig was successful in recording a two-minute, two-way conversation with this same friend. Several people were present while this took place. The deceased friend gave his first and last name and everyone recognized his voice, as it sounded exactly the same as his voice while in life.

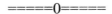

Betty's father had died in 1948. In 1984 Mercedes Shepanek began recording EVP messages from him. She sent Betty a tape of her father's voice saying, *"This is Illtyd Evans speaking. This is Illtyd!"* In 1988 EVP researcher Clara Laughlin recorded a voice saying, *"This is Illtyd, this is Illtyd."* When the two tapes were compared the voice Clara recorded sounded exactly like the voice taped by Mercedes in 1984.

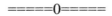

Erland Babcock speaks of a time when he was close to being an atheist and how his thoughts have changed. He had a friend fall off a roof, break his neck and die. Erland was able to contact him through EVP and the friend said, *"Everything so beautiful ... there is a God!"*

The mother of Lee had a miscarriage some years ago. When Lee asked her deceased grandparents to speak, she often recorded a young boy's voice calling, *"Grandpa."* She could not understand this, as there were no grandchildren who had died. Finally, Lee asked her grandparents if the baby boy who was with them was the baby that her mother had miscarried, and if so, what was his name. The grandmother replied, *"Why don't you ask him?"* On her next recording she did this and a young boy replied, *"Evan."*

=====0=====

Alexander MacRae, the developer of the Alpha Device[30] for EVP, tells of some interesting things that happened in the early days, when he was conducting presentations of the Alpha in an effort to find backers to fund further research.

At the Buchanan Street Hotel in Glasgow, he conducted a demonstration for two physicists and a young lady. One man claimed he heard his late wife's voice. Later, the man told Alec that he had been so depressed; he was on the point of suicide. Hearing the EVP changed that and he was so impressed that he set up another meeting for Alec to present the Alpha in the boardroom of a major Scottish electronics company.

Present at this meeting were the two physicists, the financial director and a company engineer. On playback, an elderly lady's voice called out, *"Andy!"* The young engineer's face became red and he left the room. Later, the widower whose name was Archie, said that he had heard a message from his wife.

Alec returned home and analyzed the tape but had to tell Archie that the voice was not his wife Margaret but someone named Molly. "Oh," the widower said, "It was my wife Margaret, Molly was just the pet name I had for her."

At a demonstration of the Alpha Device by Alec to twenty visiting scientists at the Palace Hotel in Inverness, Professor Leslie W. said she heard her brother's voice. Alec later isolated the EVP to hear, *"Leslie. This is it. David."* After listening carefully he detected that the first name had three syllables like "Lessillee." When Alec admitted this to the professor, she said, "God bless your good ears Alec. That was what they used to call me as a child!"

They are Aware of Us, and What We are Doing

At one time or another, every experimenter will realize that those on the other side can see us and they can see what we are doing. It may be their comment about a new piece of equipment we are trying or about what we are wearing. Both of these things have happened to us. They probably do not watch us all the time or invade our privacy. However, there are others, like Charli Claypool, who feel that her spirit friends are always there and even go with her when she travels to craft shows. On her recordings, the spirits talk about everything in her life, from making lemonade to what she is preparing in the kitchen.

When our cat has entered the room while we were conducting an experiment, we recorded an EVP commenting on how lovely our cat is. They have also commented on our activities that are away from the experiment room, such as the fact that we are preparing for a trip. During one of our early recordings that we will never forget, we recorded our first direct voice from a radio that we could hear in real-time. What it said was certainly truthful and also amusing.

The recording setup at that time consisted of a foreign language tape that we played for background noise. Our bedroom was across the hallway from the room we used for experiments, and we had an old clock-radio next to the bed. A few odd things had happened with the radio in the past, so we turned it on as well for background noise during experiments. It was tuned off-station to white noise and the volume turned up so that we could hear the static from the experiment room.

During one interesting experiment, we were more tired than usual and ended the recording early by saying, "We have some things to do and so are going to end the session a little early." A loud voice boomed from the radio by the bed saying, *"You want to watch TV."*

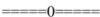

Jacques Blanc-Garin's first wife, Annick, crossed to the other side. Several months later Jacques had a car accident in snow and black ice, in which his car went over a protective rail. He was all right and could not understand how it had been possible to come out of the accident unhurt, as his car was destroyed.

Monique Laage a good friend of Jacques who later became his wife, had called another friend, Genevieve, to tell her about the accident. Genevieve had been working in EVP for quite a while and she was recording during the phone call. Genevieve told Monique, "It's a small miracle that Jacques didn't get hurt." When Genevieve listened to the tape, a clear woman's voice said, *"But Monique, I was there."* When Jacques listened to the recording he perfectly recognized the

voice and intonations of his deceased wife, Annick. He then understood that she had been there and protected him during the accident.

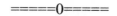

J.L. sat down at her recorder with a red hat on, something that she had
never done before. She later found this message on her recording,
"This is Alvin. I love you. I love your hat." Alvin was the boyfriend
she had as a young girl. He had died four years before.

Learning Lessons about Communication through EVP

When we communicate with our friends across the veil, we should not
forget that they are still much like us and should be treated ethically
with the respect and consideration that we would have for any of our
good friends who live in the Physical Plane. These paranormal voices
can seem so fantastic when we first discover them that it is easy to
forget that our contacts are indeed real people, just like us.

You may remember that we lived in Kansas when we first began
recording EVP. At that time, the Midwest was hit with intense rains
that lasted weeks; rivers flooded and levees broke. The news was filled
with the pictures of people fleeing their homes, and then with the pictures of the animals and livestock that were left behind.

Lisa was heartsick over the plight of these animals and immediately
began asking our contacts on the other side about them. The entities
told us that the *"animals also survive."* They told us that many on the
other side work to rescue the animals and help bring them to the other
side. They said, *"The pictures are sad,"* but that *"they continue"* and
"nothing that lives there is ever lost."

Each night, we continued to ask about the animals in our recordings
and our contacts continued to reassure us that the animals were being
helped and taken care of. It was about the seventh night that Lisa once
again asked for a report on the animals. A loud frustrated voice replied, *"Stop with the animals! They are all right!"*

This led us to realize that we had become as irritating as a two year
old who incessantly repeated the same question. They had certainly
been as patient as any parent might be with a two year old, but obviously, a two year old can get on one's nerves. Thus, we learned an important lesson.

=====0=====

David Sircom reached an individual named Jennifer in his EVP recordings. He first wrote about this to us saying, "A few months ago, I
made contact with Jennifer. She gave me her last name, after two
months, but swore me to secrecy about revealing it. She has been very
helpful ... her voice seems to be getting stronger." Jennifer was
David's first contact to give her name, age and when she died. She
told him that she had been a teacher in life.

What happened next was very intriguing. A very ominous, growl-
ing voice in an EVP warned David to leave his wife (Jennifer) alone.
David wrote, "Well of course I didn't until she actually told me it
would be better to stop 'asking' for her. She sounded so sad, almost
crying. Not wishing to cause her any additional grief, I stopped asking
for her."

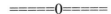

Sarah Estep has written about an incident in which she heard the name
of a well-known scientist on one of her recordings and she called on
him frequently during the following week. Messages came through on
both sides of her tape indicating that he was with her and also bringing
other scientists in to give messages. After a week she recorded this
message after calling on him, *That's still my name but I wish she'd
never heard. I rest!"* When she heard this she apologized and told him
that she did not mean to bother him and that she would not call on him
again.

Sarah wondered why this person could not just walk away and ignore her. She decided that he or one of his colleagues had spoken his
name and that she had not been meant to hear it. She wrote, "Once I
had come upon his particular energies, his frequencies, he found it
hard to break free. It was almost as if I had to 'release' him, which I
did as soon as I knew how he felt."

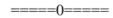

Talking about "release," those who pursue getting paranormal voices,
and succeed, will inevitably find that they will at one point or another
record voices asking for help. Our nonphysical friends probably cannot be released to move on unless they want to, but when you record
voices that ask for help, take just a little time to try to find out more
about them and how they need help. You might tell them that they are

in another dimension from you, and that often, the people that you have heard from in the dimension that they are in now, have departed from earth life. Tell them they are very much alive but they are vibrating at a different frequency and that they need to find the untold numbers of souls who live in their higher plane of existence. Tell them that you are sure there will be a light, and if they go toward it, they will see loved ones that they know.

You should receive an interesting response if this person is from another dimension and is not someone who has died and is confused. Many researchers record voices from what we would call extraterrestrials and from others that have never lived in the Physical Plane. But nine times out of ten, an entity who asks for help is confused, frightened and does not know that there is life after death. Some may not know what to do. Others may simply not even realize that they are dead. Some spirits are afraid to go on because they have been taught that there is a hell.

Those on the other side have told us that there is no hell and that there is no one that will judge another person's life. The individual entity will do the judging of the life it has led. There is constant progression and learning, not only here but in the next dimension as well. If the life was not what the entity wished it had led, it will be able to make up for and learn from past mistakes. The only possibility of hell is if we should create it for ourselves. Thought creates reality in this world and even more quickly in the next. Our spirit communicators have also told us that many on their side try to reach these "stuck" individuals, but that they are often vibrating at a level closer to our own and are difficult for them to reach. It is the people still living in the physical, they say, who have a better chance of reaching many of those who are stuck close to the Earth Plane.

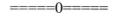

A.R. recorded a message from a man saying, *"I'm from another world. Can't get help. Won't you help me?"* After receiving the message, A.R. prayed for the man, told him to look for a light and follow it, and as he did so, to ask for help. Several days later, she heard from the same man who said, *"I'm telling you they've gone free. Got a message set them free. It was quite a scene!"* The entity spoke again three days later saying, *"You should see our school. It's out on farmland."* The man told A.R. that he would be at the farm for six months.

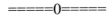

It takes two to communicate and if the entity we are trying to reach on the other side for some reason either can not or does not want to participate, communication is not going to happen. Bill Weber had an interesting experience in which an entity spoke, but made it clear that this comment would be it for the time being. He had conducted a recording for a friend whose brother had committed suicide, leaving his parents, wife and child behind. Bill called on the man without receiving a response and then called on him again. After ten seconds a loud clear voice said, *"Nobody feels like this. Bye Bye."*

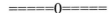

Olga recorded a clear voice that said, *"Olga you are protected. I will love you. We will see each other again."* During a later recording, she commented that it must be difficult for people who die because they are still able to perceive their loved ones, but are not perceived in return. A loud voice came from her radio and said, *"That is true!"*

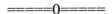

A communicator from the other side gave some advice to the famous researcher, Hans-Otto Koenig, and others in a recording made in Germany. The message was, *"Love and patience are very important for your contacts.... A stream of energy [helps] to build a bridge with earth.... Your loved ones be not alone but are of the same vibration within the energy field.... Damage to the material body has no influence on the astral body."*

Do Our Pets and other Animals Survive?

You have already been told you about how we learned early on that all animals do survive. Here is another story regarding animals that represents a personal breakthrough for us.

It is clear that we were fortunate to be able to record EVP voices so quickly. Lisa heard that first EVP, *"Crystals Help,"* on the third evening and the EVP have continued to come through since. One of the frustrating things about the EVP we were collecting was that, in the beginning, the voices were low and very much in the noise. Lisa was able to hear them but Tom could not. This left Lisa more than a little uncertain as to whether or not she was truly hearing phenomenal voices in the white noise. But this too had a breakthrough.

A couple of months after the first voices came through, Lisa had an experience with a nest of birds near her office building. She was a Facilities Manager and managed several large buildings for a major corporation. Each building had a Building Manager, with whom she would coordinate change orders. One of these managers had become a good friend. The two were outside one of the buildings one day and they noticed a noisy nest of baby birds. After that, they went out each day to check on the progress of the babies, waiting to see them fly.

The day before a three-day weekend, they noticed that one of the baby birds had fallen out of the nest. The parents were continuing to feed it on the ground and the two thought that perhaps it was learning how to fly. Both were shocked and saddened when they returned from their holiday to find the baby bird dead. Lisa was very upset that they had not tried to put it back in the nest.

That evening she told her friends on the other side how sorry she was for not taking action to save the baby bird. She asked, "What can I do now that the baby bird is on your side?" A very loud, lovely and airy female voice said, *"Release and remember."* This voice was so beautiful that it sounded angelic. She was also an angel because Tom clearly heard her voice and was able to hear EVP messages from then on. It was not just our first Class A EVP. The message was profound and has stayed in our memories. In fact, "Release and Remember" was the original working title for this book.

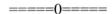

Sarah Estep's husband, Charlie, transitioned to the other side February 15, 1998. We asked Sarah if Charlie had communicated with her through EVP about their pets.

Sarah wrote, "On February 15, 2002, which was the fourth anniversary of his death, I devoted the recording to him. I do this each year at that time and often have a response from him. This last February, I was ending the five-minute recording and said, 'We gave each other three wonderful children.' Immediately, in conversational style that made me feel as if he'd been sitting beside me during the five minutes, he replied, *'And there's Misty.'* ...

"Misty was the ... last dog Charlie and I had together. ... Several months after Charlie passed on, he came to my recording room and said, *'I brought Misty down with me.'* When I asked him in the next recording what he and Misty did, he replied, still Class A, *'We play.'*"

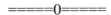

Carol Barron-Karajohn knows that our pets can respond from the other side. She wrote, "Shortly after my cat Bambi died, I taped a female voice saying, *'Bambi.'* During another session [on another night] I heard a cat meow upon playback."

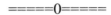

J.E., of Germany, reported that her dear dog, Rolf, died, and that this was very sad for her as it is for all of us who love our pets as if they were our children. She frequently asked about the dog when she recorded and often received answers. On two or three occasions, he even barked for her. Three weeks after his death she recorded, *"I already am friends with the dog."*

She thought that she would never get another dog because her grief was so great, but two years later, she read about a man who was giving away dogs and went to see him. Two dogs appealed to her and she adopted them. She named one Wanda. Wanda had many of the unique behavior patterns that Rolf had and knew many things that only Rolf knew. When she went to a national ITC conference in her country, she took the dogs. She entered a room with two hundred people, and even though both dogs were shy, Wanda immediately ran across the room to a man who had been friends with her dog, Rolf. Wanda climbed onto the man's lap and licked his face. Both J.E. and another friend received taped messages indicating that Wanda was once Rolf.

Gerda Slater had to put to sleep her well loved cat, Shamballa. While taping she asked her husband and parents, now on the other side, where Shamballa was. Her husband, Ed, answered clearly, *"We have him!"* This was followed by three meows. A month later she asked Ed if Shamballa was with Nicky, their Schnauzer dog. The two had played together before Nicky's transition. Ed answered back loud and clear, *"Yes!"* This was followed by several loud, shrill barks.

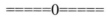

I.M. and her father shared the love of a little dog named Boopie. Her father died and then Boopie died a few years later. After the dog's death, she was recording in her kitchen. The tape had several messages, first the word, *"Jump!"* came through and just a few seconds later, *"Boopie!"* followed by her father's voice saying, *"I'm taking care of Boopi."*

=====0=====

Martha Copeland has been communicating through EVP with her daughter, Cathy. Martha and her sister were conducting a recording and picked up Cathy's voice saying, *"Mama, I have Muffin."* This was followed by the sound of a dog. Muffin, had been Cathy's dog for seventeen years, and had transitioned shortly after Cathy. Martha has heard Cathy calling for her dogs, Shishi and Gretel, or talking about the other animals in Martha's house.

Before Cathy crossed to the other side, she also had a pet rat named Elainey. Martha's sister, Ginny, took the rat because Martha was, although a true animal lover, not keen on rats. Ginny, on the other hand, had become very fond of the rat.

One day, Ginny was using her daughter Rachel's computer. Rachel was sitting on the bed studying for an exam. Ginny made an EVP recording using the computer, and when she played it back, they could both hear Cathy singing, *"Elainey, Elainey, I miss my rat, Elainey!"* Martha suggested that Cathy might be preparing Ginny for the possibility that Elainey might soon transition. Weeks later, Ginny recorded an EVP indicating that Cathy was coming for Elainey. Elainey transitioned not long after that.

Messages Showing Cross-Correspondence

Twenty years ago, Sarah Estep recorded dozens of messages in which those on the other side requested that she place a mirror in her recording room, so she went to a local variety store and bought a mirror. After placing the mirror on top of her reel-to-reel recorder, she received seven or eight messages regarding the mirror, including *"Sarah has a mirror,"* and *"We can talk into it!"*

Three to four weeks after Sarah began using the mirror, another AA-EVP member and good friend called and said that he had been recording messages telling him that he should use a mirror. This friend did not know that Sarah had received mirror messages as well.

A few weeks later, AA-EVP member, Mercedes Shepanek, called Sarah to talk about mirror messages. Mercedes was also unaware of the mirror messages that Sarah and the other member had received.

She told Sarah that she had been receiving many messages telling her to use a mirror, and so, she had purchased two mirrors and had placed them on either side of her reel-to-reel recorder. The entities seemed very happy with this and told her that, *"Two mirrors are enough!"*

After she had read the draft of this book, Sarah wrote to us, pointing out that, "Alexander MacRae of Scotland also reported receiving messages telling him to get a mirror for his experiment room. A woman in England reported the same thing. None of us knew at the time anyone else had received mirror messages."

Sarah did not feel that this series of events was a coincidence. She felt that, "Someone or something in another dimension was showing all of us at the same time, that they were aware of us, and wanted to let us know. They accomplished this by giving us almost identical messages, proving that what we got on tape was coming from an unseen dimension.

"Are mirrors vital to communication?" Sarah questions. "I don't think so. Most people who tape, and have good results, do so without mirrors.... I still feel the main reason for the mirror messages ... was to present us with a synchronistic experience."

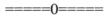

Clara Laughlin went on a trip with her daughter and a girlfriend for a few days. Without reservations, they found that there was only one place that could give them a room and that would only be for one night. The man at the counter held out little hope that a cancellation was possible.

The room was lovely and Clara said out loud to Tom, her husband now on the other side, "For heaven's sake, do something so we won't have to move. There is nothing else." The next morning, Tom came to Clara in a dream. She was standing on the first floor in front of the elevator. Tom was there and looked her straight in the eyes and handed her a key that was exactly like their room key without saying a word. The next day, the front desk informed them that there had been a cancellation and that they could keep the room.

When Clara arrived home she made a recording and asked if anyone had been with them on the trip. You can imagine how Clara felt when she played her recording back and heard, *"Elevator. Key to my wife, Clara. Tom."*

=====0=====

Ann Longmore-Etheridge wrote us about an interesting case. In the late 1980s, her mother was a food service worker at an elementary school. She was not fond of many of the students, but she did tell Ann some amusing stories about a little girl named Rosie. Her mother described Rosie as a friendly dark-haired child who was half Asian. One day, Ann's mother tearfully told her that Rosie had been kidnapped.

Ann has been psychic since she was a child, and had communicated with discarnate entities through automatic writing, as well as clairvoyantly. At the time of Rosie's abduction, she had grown skeptical of her psychic abilities and would not accept her psychic impressions as evidence of life after death. She wanted and searched for incontrovertible evidence of survival. Ann read Sarah Estep's book and turned to experimenting with EVP. She felt that if she could record voices that were strong and clear, and that were in answer to questions that she asked, she would then be convinced of after-death survival. At the time Rosie disappeared, she had been taping with limited success for six months, but had not yet received a loud, clear voice and had become frustrated with the faint whispers or voices that she was recording.

Out of sympathy for her mother, and despite her disenchantment with her psychic abilities, Ann decided to ask her spirit friends to look over Rosie's situation and was told through automatic writing that Rosie was dead and was now in the care of a woman who may have been a relative. Her body was found in a ditch the next day. She had been strangled. Police later tied her murder to a string of killings by a sexual predator.

Within a few days of her death, Rosie herself put in an appearance through Ann's automatic writing, asking her to tell her parents that she was still alive. Ann wrote, "I explained that I couldn't tell them because they would think I was crazy. (I'm sure I looked crazy arguing with a black space in my living room.) I felt that Rosie was frustrated and I believe that what followed was her attempt to convince me to reconsider.

"I had never heard a child's voice among the EVP I had received. The first time I did was soon thereafter. I had announced that I was about to turn off the tape recorder and a young girl's voice replied in a clearly understood voice, *'I knew it.'* Then, after a beat, *'Goodbye!'*

Not many days later, I taped what sounded like a room full of people talking. I thought I could pick out a child's voice saying, *'That's amazing Spanish, Harry.'* This thrilled me because I did, in fact, frequently call on an entity named Harry. I also often asked to hear from another spirit named Francis. Several days later, I was taping with a friend. I called on Francis and when we played the tape back, we were startled by a voice that was louder than my own. We were so frightened by the unexpected volume that we clung to each other, shaking. It was a little girl calling, *'Francis!'* as if she summoned him to speak. I would hear from that little girl again soon, although not by the mechanism of the tape recorder."

Ann's family has been part of the congregation of a Spiritualist Church[9] for multiple generations. A few months after she received this loud voice, her church was playing host to the Spiritualist's annual conference. One night during the event, and at the end of a special church service, a crowd had gathered to hear four respected mediums offer spirit greetings to the congregation. Ann told us that, "Because of time constraints and the huge audience, it was announced that each medium would give only about a half-dozen messages; therefore, out of a crowd of about two hundred and fifty, around twenty-four people were going to receive a message from a departed loved one.

"I had arrived late and was sitting in the very last row of pews, in a dark corner. Despite this disadvantage, the first medium stood up and pointed to me. He began by mentioning a long list of people who were there for me, including my beloved Uncle Carl and Aunt Helen. As I was trying to commit to memory all these names, and marveling over my relation's appearance, the medium described a dark-haired girl with oriental eyes who was out in front, demanding that I acknowledge her. In a moment of what I've heard the medium, John Edward, call 'psychic amnesia,' I drew a blank and shook my head. 'She says you know her,' the medium stressed. I just shook my head. Seeming quite frustrated, he told me to remember what he'd said, and then he went on to give other messages.

"When he had finished, the second medium took over. She was the Reverend Anne Gehman, who participated in *The Afterlife Experiments, Breakthrough Scientific Evidence of Life After Death*, as chronicled by Gary E. Schwartz with William L. Simon.[71] She stood up, swept across the podium in her long white gown and pointed straight at me. 'Who is ROSE?' she demanded, jabbing her finger at me to stress each word. 'She's telling you to recognize her. WHO IS ROSE?'

"I remember my mouth falling open and saying something to acknowledge my sudden understanding. Gehman nodded and said, 'Good, because she's not going to give up until you get it.'

"Although I felt Rosie's presence around me in the months that followed, I never heard from her on tape again. In time, she seemed to move away into the new world that was her home. Sadly, I still failed to work up the courage to give Rosie's parents her message. I often feel bad about that now, because Rosie gave me the best evidence of survival that I have ever had—proof that I could not ignore or rationalize away. She brought back my hope, at a time when it was low. I wish I had done half as much for her."

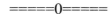

You have seen many EVP examples that have come from Martha Copeland's communication with her daughter Cathy. Another AA-EVP member, Karen Mossey had lost her son Rob. Karen wrote, "I cried reading Martha's Emails on the AA-EVP Egroup. I wanted to reach out to her and there was a driving force that I had to know about Cathy. It was stronger than just emailing each other. We began to talk, finding we shared so much in common. We found that our children were in fact very much alike."

Karen asked her son Rob to send her a dream. He did send a dream, but not to Karen. Instead, another of Karen's friends, Judy, who did not know either Cathy or Rob, had a dream about Rob, as did Martha.

Martha's daughter, Cathy, came to her in a dream and told her that she had a new friend named Rob and that he liked to fish all of the time. In her dreams, Martha had normally seen Cathy in a meadow near the ocean. However, with the news of Cathy's new friend, Martha began seeing her in a log cabin. She just could not understand this, as Cathy was very much connected to the ocean before her transition.

As Judy related her dream to Karen, "Karen, your son came to me in a dream last night. I have never met him but I know it was him." Judy said that, in her dream, Karen and Judy were managing a small store with another friend. Karen and the friend had to leave the store and the minute that they did, Judy saw a bright flash of light and then Karen's son, Rob, appeared with a young woman. Judy said that the girl was Rob's girlfriend and she had blonde hair. (Cathy has blonde hair.) Rob told Judy, "Tell my mother I am happy and that I have found a new love."

Karen wrote, "This was such a confirmation for Martha and I that Rob and Cathy would come through together to a stranger, my friend Judy, who did not know either one of them. Judy believes that I am still too fragile to receive Rob in person and that is why I needed to be taken out of the store and why the message had to come through her."

Karen continued, "I immediately called Martha because she had just told me about her dream in which Rob and Cathy were in the fishing lodge and Rob was cooking fish. He was always cooking fish! Rob's passion was fishing." Karen told us that Rob's headstone even has a picture of him fishing, with Karen and the rest of the family having a picnic. "Gone fishin" is etched in the black marble.

As even further proof, one night Karen called Martha and told her that she had remembered something. She had been given a little log cabin bird house when Rob died. She sent a picture of it to Martha and Martha immediately recognized it as the log cabin in her dreams.

Karen next began to have a repeated vision of Rob and Cathy dancing, which she felt that she needed to paint. Martha and Cathy loved dancing. Rob, on the other hand, was not very fond of dancing. A week later, Martha called Karen and said that she had another dream. Rob and Cathy were in the log cabin again. Rob was sitting with two of his friends. (Karen wrote that, "Ironically Rob had two very good friends, Brandon and John, who passed on before Rob—as with Cathy and Rob, because of auto accidents.") In Martha's dream, the boys were watching television. Also in the dream, Cathy told Martha that Rob had sent the vision of them dancing to Karen to let her know that he is happy. Rob also told Cathy in the dream that, "My Mom is an awesome artist." These are the exact words he used to tell Karen this, before his transition.

Martha told us, "I had another strange thing happen regarding Cathy with Karen. Karen had wanted to paint a picture of Cathy and Rob dancing. Cathy had a very nice wooden artist studio box that had been missing for some time and I wanted to give it to Karen." Martha had repeatedly looked in Cathy's room but the box was nowhere to be found. She wrote, "I kept hearing Cathy's voice in my head telling me to look in her room. She seemed to be saying, 'Mom look in my room one more time.' When I did, I found the artist set placed in the middle of her bed. I guess Cathy wanted the box sent to Karen, too!"

Karen is also getting EVP messages about Rob but is not yet sure if the EVP are coming directly from him. Karen wrote that, "One message lately was, *'Robbie, you're needed here,'* so somebody was talking to him."

The two feel that their meeting has been a miracle that was set into motion, and very much meant to be, through their membership in the Association, and Cathy and Rob working together from the other side, to make it happen.

Using EVP for Grief Management

Linda Williamson sits with a group of woman who are using EVP to assist others with overcoming the grief of losing a loved one. Recently Linda worked with a woman who had lost her daughter twenty years ago. The woman had not gotten over this death and was angry with God for taking her child. Linda was able to help this woman through an EVP message from her daughter and the woman was overcome with joy.

Linda has also been using the telephone to capture EVP with great success. She finds EVP voices in the messages left by friends on her answering machine, and with permission, she also records telephone calls that she has with people. EVP messages from the caller's loved ones are always found.

Larry Dean left a message on Linda's answering machine and then completed a call to her a few days later. At that time, Linda told Larry that she had found EVP messages on the answering machine recording. She then sent the results to Larry, who was amazed. An EVP saying, *"Jim,"* was on the tape several times. This is the name of Larry's father and grandfather. Another EVP clearly said, *"This is Edison."* Larry and his fellow researcher, Patricia Begley, have been told that Edison is one of the scientists working with them to establish ITC contacts.

Linda recorded a subsequent phone call and received an EVP saying, *"I love you."* Patricia immediately recognized the voice as that of her father who had also come in on one of the couple's ITC experiments.

Tina Laurent wrote to us about a tape that Linda had sent to her. Linda had recorded a telephone conversation between the two. Tina said that there were nearly forty EVP utterances, and although some of them were indistinct, most of them were quite loud. The word, *"Tina,"* was spoken frequently. Tina recognized the voice of her first husband saying, *"I'm Helm."* Tina has received EVP messages from him on several occasions and is familiar with his voice. The word, *"Mona,"* was also received and this is the name that Helm had called Tina when he was on earth. One of her favorite uncles also came through giving his name, *"Horace."*

Linda is deeply committed to using EVP to help people through the grieving process. Her team of spirit helpers has a great ability to do just that by bringing in loved ones from the other side.

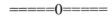

Susan Bové lost her Aunt Margie after years of illness. Just three months later, her Uncle Tony, to whom her Aunt had been married for fifty-seven years, was dying of cancer. Susan's cousin was desperate for confirmation that her grandmother was nearby and asked Susan's sister to place a recorder on his pillow to see if any voices could be captured. The day that he died, they recorded for forty-five minutes and captured six spirit voices. Susan wrote, "While I listened to the recording for voices, I immediately started crying when I heard the voice of my Aunt Marge! It was so clearly her, as she sounded toward the end of her own life." The recording has been passed around their large family and everyone has been astonished that it was her Aunt's voice. One particular EVP voice that they treasure can be heard saying, *"I love you."*

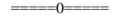

Tina Laurent previously gave readings for people. She customarily used her tape recorder the day before a client was to see her, to see what messages she could record for them via EVP. She often recorded useful information and would play the voices and messages that she received for the people when they came to see her. Many were very evidential.

On one particular recording, she asked about a client who would be coming the following day. A man's voice replied, *"Uncle Tom."* He then went on to call her his *"little girl"* and said that she had *"a big decision to make."* Tina did not feel comfortable with the message as hardly anyone in Wales is named Tom. She did, however, go ahead and play the messages for her client. The woman began crying. She did have an Uncle Tom who had been very close before his death and she also admitted that she did have a big decision to make.

On another occasion, Tina was recording with a friend. The name, *"Max Wall,"* was recorded, which was a name that was unknown to Tina. The friend became very excited after hearing the recording, as this was the name of a man that she almost married many years before. He had died at the age of thirty-three.

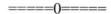

A French researcher, Monique Simonet, was recording for a friend who had just lost her husband. As Monique was soon to learn, the husband had not been a very good man. As soon as she started the tape, the husband came through saying, *"I was waiting for your call. I love you. Please forgive me. I am waiting here for you."* After these messages, he was quiet, and another voice came in and said, *"Contact ended. Pray for him."* After the messages came through, Monique's friend began crying. In confidence, she told Monique things about her husband that were shocking and that Monique had known nothing about.

Messages of Wisdom from the Other Side

The transcommunication group called the Mitmenchlicher Transcom-munikations Forschungs Dienst (MTFD), formed in Frankfurt Germany in 1987, received a video and audio communication in January 1988. Dr. Ernst Senkowski was present at the time. A male voice was heard over radio and his face was seen in black and white on the television set. He said, *"The whole consists of good and bad. Your existence is made up of every type of feeling. Every feeling is reality and you are responsible for everything. This is how your being (life) gains purpose.... By your own energies you create good and evil."*[54(V9N2)]

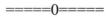

On March 23,1990, Manfred Boden, who is well-known for his contacts with those in other dimensions, made his transition. On the day of his funeral, March 29, he communicated through the computer of the MTDF group in Germany. The communication read, *"I am in a famil-*

iar world of unlimited diversity. Everything is possible. Middle plane of existence. The Past, Present and Future join logically together. I am seeking experience as my cycle of reincarnation is finished. There are infinitely many systems. It is as if I am experiencing everything in a dream. There are a great number of people here who are confused. Everything is part of the all. I need rest and concentration."[54(V9N3)]

Boden gave evidence to prove his survival. The word, *"Charly,"* was given. This was the identification he used as part of a Citizens Band radio system during his life. In addition, Boden and Fritz Malkhoff had agreed on a code word, "L", and this also came through from him.[54(V9N4)]

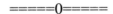

In an early communication to the Harsch-Fischbachs at the end of 1986, Technician said, *"The grief and suffering people bear and have to go through is a part of their inner self. Some of it is through their own action or initiated by higher forces in order to activate the learning process that leads to recognition, improvement, and perfection... It is all closely connected and tied in with free will and choice of the individual, which God's power has granted to each of us as a great gift.... Without free will and choice, there is no recognition of truth which comes from within. Therefore, blind obedience is not what higher powers want.... God prefers the seeker and those who question. No efforts are spared to advance human thinking and individual initiative from its low animal instincts to a position of spiritual thinking."*[3(167)]

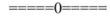

In 1992, Konstantin Raudive made a phone call from the other side to Adolf Homes. He first confirmed the contact made by Doc Müller to Homes via a small radio. He then went on to say, *"All your daily events, your thoughts and actions as well as events you seem unable to control, such as environmental catastrophes, in the final analysis were created by yourself, though they originate in other dimensions, in dreams and in other, trance-like conditions when you were your real self.*

"None of you has to die in order to be what you would like to be. You only have to die, because you selected one of many possibilities that were open to you. Even the person who is panic-stricken for fear

of death, long ago in another dimension decided his physical death."54(V12N1)

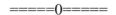

In 1994 Hans Bender, former parapsychologist of the University of Freidburg, Germany, spoke through the radio in the home of Adolf Homes.54(V13N3) He said, *"Your system of reality is one of countless others. All are happening at the same time. The 'frequency' of our own reality is so short that it cannot be perceived by you. It is far outside the range of electromagnetic wavelengths and has nothing to do with it anymore. Do not visualize that we exist about you such as in 'Heaven.' The concepts 'Above' and 'Below' are products of your mind. The soul does not swing upwards, it exists in the center and orients itself in every direction. Things which you create with your mind are always part of your post mortal life whether they seem real or not. This is also true of your religion. You shall always find what you created in your mind such as a benevolent God or an evil Devil. Therefore, concentrate on the depth of your consciousness and on what you consider positive and good."*

As new experimenters and researchers enter the field of EVP and ITC they often ask the same question, "Why do those on the other side not tell us how to improve our equipment so that better contacts can be made?" Maggy and Jules Harsch-Fischbach often noticed that their technical questions were put off as not urgent, and that their spirit friends preferred questions of human interest. They began to realize that contacts through their equipment were not triggered by intricate technical circuitry and its understanding but by a state of inner readiness, efforts for a positive spiritual attitude and many other factors that they knew little about. They wrote that the words of technical advice that they did get from the other side were merely hints and that they did not know why a certain piece of equipment would work well when used by one experimenter and not another.

Technology is an important part of EVP and ITC, but as you can see in these examples, transcommunication is between people and is very much about human concerns. From these messages, we know that the proof of their validity is in the comfort and guidance they offer those who have sought to communicate with loved ones.

Chapter 7

Ghost Hunting

Most EVP experimentation is accomplished in the experimenter's home under controlled conditions. If you are ready to record EVP in this fashion you can go to Chapter 10 of this book and learn how to begin.

There are a growing number of people who like to conduct EVP experiments in the field, especially in locations that are thought to be frequented by discarnate entities. In principle, the techniques are the same for both situations, except that the environment or the recording equipment must provide the necessary background noise in field recording.

There is evidence that EVP collected under field conditions may be initiated by entities who are in very different circumstances than those who are called on under controlled conditions. This may be because of the expectations of the experimenter and the way the session is conducted. It is clear with recording in the field that we expect to record EVP from the resident ghost. The evidence is accumulating in support of the belief that some locations, such as sacred locations and old buildings, retain the energy of past events, and sometimes, the spirits of the people who participated in those events. There is also growing evidence that certain geophysical conditions, places and buildings that have had long-time human occupancy may cause the formation of etheric vortices through which entities may more readily visit our world. As such, some entities encountered in field conditions may not be "local" at all, but may be simply there for a visit. It is also possible that those who record in controlled conditions may have some of the members of their own spirit team go along with them on field trips to haunted sites.

In this chapter, we will address the differences between EVP collection in controlled circumstances and EVP collection in the field. This will include techniques for field recording and some of the special considerations those recording conditions suggest. If you are planning to collect EVP in the field, we recommend that you take time to visit some of the paranormal or haunting investigation websites to see what they have to say about the subject. Many of these sites have excellent information with examples and "how to" instructions. From members in the AA-EVP who belong to various investigation groups, we know that there are many different ways in which an investigation may be accomplished.

Differences Between Controlled and Field Conditions for EVP Collection

Stating that EVP or ITC experimentation is best accomplished one way or another is usually not a good idea, since the results of a particular technique will differ depending on the circumstances. Every situation is different and every experimenter brings unique characteristics to the experimental situation. As a result, EVP messages may be more or less meaningful, evidential or clearly understood depending on the experimental circumstance. Our comments here are designed to reflect the trends that we have observed. You can expect that your actual experience with these phenomena will be somewhat different.

In experiments conducted in the controlled conditions of the home, the experimenter usually asks questions or asks for a specific person. Also, controlled experimental sessions over a long period of time appear to attract entities that are often willing to assist the experimenter in improving communication. Thus, the resulting EVP tend to be more uplifting, and are sometimes, clearly initiated by entities who have developed a working relationship with the experimenter or who have been asked to speak. The quality of EVP collected in controlled situations is not necessarily greater than EVP collected in field conditions, but we have noticed more frequent reports of recognized voices and answered questions from people who record in their home using common recording techniques. On the other hand, we feel that it is easier to get EVP voices in a haunted location.

In field recording, EVP experimenters are usually seeking to record comments from the entities that are thought to reside in that particular location. The experimenter does not expect to hear from a loved one, but does expect to hear from whichever entity that might be responsible for the reports of possible haunting activity in the area. Thus, the

EVP from field conditions do tend to be consistent with an entity that might be "earthbound" or "stuck" in the haunted location. Sometimes, these are angry or confused sounding EVP, such as, *"Get out!"* or *"Turn it off!"* But often they are the voices of friendly entities wanting to be heard and saying things like *"Listen to me,"* or giving helpful information like a name or telling about some event that has taken place in the location. A common type of message, and one that we have collected, is the communicating entity saying something like, *"What is happening?" "They see us"* and even *"help me,"* indicating that the entities who are present are curious and can see what the investigator is doing.

Through our own experiences with field recording, we have learned that not all EVP collected in a haunted location will be from the resident ghosts. At first this was just an idea we had, but the entities proved it to us. On a trip to Alcatraz Island in the San Francisco Bay, we recorded many messages that fit the location; however, other voices were recorded that seemed enlightened. These "enlightened" voices perplexed us. Could a being that had not gone on, in other words a ghost, be profound?

In the Alcatraz prison, we paused a moment to admire the view through the heavily screened window. Alcatraz was first a military prison and then later a federal prison. The view from the window was very beautiful. To be locked up on an island with beautiful scenes from every window must have been living hell, and we said this. The recorders were set to "Voice Activation," and on one recorder, there is a loud clear distinguished sounding voice that speaks immediately after our statement saying, *"Think positive, instantly away."* This seemed very profound for a "stuck" entity.

It was not long after this that we recorded an EVP that proves that entities who are not "local" will come to us, even on field trips to haunted locations. This was very clearly demonstrated when we were in the hospital area and in the hydrotherapy room of the prison. This is the room that many psychics have said contains a great deal of negative energy. It was in this room that we received an EVP from a group of ladies singing the name, *"Paolo Presi."* Paolo Presi is a well-known ITC researcher in Italy. Immediately after this, the Italian word "Bongiorno" is spoken; Bongiorno is "good day" in English. There have been several articles about Presi's work in the AA-EVP News-

Journal and we have communicated with him a few times through email. However, we certainly were not thinking about Paolo Presi at Alcatraz. It seems obvious to us that the entities used this opportunity to send a greeting to Paolo through our tape recorder. That this greeting was placed on a recording made at Alcatraz was quite a shock.

One difference between field and controlled recording is that, in controlled recording circumstances, the experimenter is able to supply the sound energy that we believe helps the entity to form the voice. The opposing viewpoints about the need to supply background sound energy will be discussed in Chapter 11, so here we will say that the evidence seems to support the belief that the communicating entity changes available sound into words. This is accomplished by a direct transfiguration of the sound, or by accumulating sound energy, and then "bursting" the energy as a phrase.

Assuming that it is necessary to have available sound energy to form EVP, then a very quiet recording device will work well, so long as sound from something like a fan or static from an electronic device is available in the room while recording. Also, communicating entities will sometimes use the experimenter's words to form their phrases, just as they use ambient noise. In controlled conditions, the experimenter is more able to keep track of voices and noises that are in the room or that come from other parts of the house.

In field recording, we have found that a very quiet tape recorder is not as effective. The solid-state digital recorders that are known as digital note takers or IC recorders tend to be better for EVP. ("IC" stands for the Integrated Circuit that contains the circuitry for this type of recorder.) One of the reasons for this, we believe, is that the electrical circuit of the IC recorders produces the necessary noise for EVP. That is, they tend to "chatter" internally as the signal is controlled by the voice activated recording switch and level thresholds in the circuitry. Some IC recorders have proven to be very poor voice recorders because of the internal noise, but for the same reason, they have proven to be excellent devices for EVP.

In field recording, it is necessary for the experimenter to have a way of keeping track of ambient voices and noise. For instance, we had difficulty finding a quiet moment to record when we were in a multi-story building and there were people talking on a lower floor.

Their voices sounded much like EVP when we listened to our recording.

Many investigators, who try to record EVP during the investigation of an active site, team up to work in pairs. As you will see in the following, it is very unlikely that any two recorders will capture the same EVP. Therefore the two recordings can be compared and mundane sounds that actually did take place at the site can be found, as they will appear on both recorders.

Another excellent solution is to have a partner videotape the recording sessions. This will provide a visual record of the session and the sound track on the video will provide a way of discounting ambient noises. To our knowledge, we have never received a documented case of two recorders picking up the same EVP. So, even though the video sound track may have EVP, using the video sound track as a control should help you distinguish between EVP and other noise made by people or just the creaking of an old building.

Hauntings Investigation Versus Ghost Hunting

This is a good time to address an important difference between experimenters who seek to record EVP in reportedly haunted locations and researchers who investigate haunting situations. EVP is fast becoming an important tool for research on haunted locations. When teamed with a clairvoyant, and instruments capable of sensing slight changes in electromagnetic radiation, heat, and movement, EVP can be very helpful in increasing the information gathered during an investigation. It can actually help describe the entity or entities that are at a particular site. The EVP voices will be male or female. The voice of a child can be distinguished and in some cases, the names of the entities present have been recorded. You may even be able to tell if an entity is angry or simply playful, and if they are earthbound or just visiting.

But true haunting investigation is much more than entity detection. It is also careful research into the history of the site and its past occupants, careful documentation of the situation, procedure and results, and if possible, an earnest attempt to help the nonphysical entity move on to higher aspects of reality. The end goal of a serious haunting investigation may be to clear the location of an unwanted nonphysical influence, and to help the occupants learn to deal with the situation. In some cases the occupants may even need counseling services. A serious investigation may add to the body of evidence that these phenomena are real, and/or add to information about the history of the haunted site.

It is our belief that the majority of EVP experimenters who like to record in known haunted locations do so out of a desire to collect EVP from the local ghost. In effect, the experimenter is a ghost hunter and makes no claim at all of being a researcher. There have been very public complaints regarding ghost groups from people who consider themselves legitimate researchers. They complain that these people are just amateurs and how such groups make a mockery of professional efforts to conduct serious research. There are ghost hunting groups all around the world. Some do make an effort to follow acceptable procedure but some are simply seeking phenomena. As long as these groups do not claim that they are something they are not, and conduct themselves in an ethical fashion, then we believe that the search to experience phenomena is a great and honorable pastime that often leads the seeker to more serious research and sometimes helps teach the truth of personal survival. It is our belief that, while some people may use the term, "haunting investigation," they actually mean something more like, "ghost hunting."

The Gold Canyon Restaurant

Janice Oberding[36] is the author of *Haunted Nevada,* Director of The Nevada Ghosts and Hauntings Research Society[37] and the Northern Nevada area representative for the American Ghost Society.[38] We met Janice at the 2002 Virginia City, Nevada Ghost Conference that she helped organize, and at which we were speakers. After the conference, we accompanied Janice and a few other hauntings enthusiasts to evaluate a restaurant in nearby Dayton as a possible site for a future investigation.

Virginia City is located in Northwestern Nevada. Janice told us that, "The town is rich in Wild West history. When the Comstock Lode was discovered, Virginia City became a haven for those seeking their fortunes. Millionaires were made and bankrupted by its silver mines, and gunslingers ruled the streets. Mark Twain began his career in Virginia City writing for the *Territorial Enterprise,* and stars of their day, such as Shakespearean actor, Edwin Booth, performed at Piper's Opera House."

Janice did not know much about the restaurant in Dayton. The owner of the restaurant had approached her at a book signing to tell her that the restaurant was quite haunted. Janice had liked her and felt that she was a credible and honest individual, and had agreed to take a preliminary look after the patrons in the restaurant had gone home.

At least for us, what took place on this preliminary look was phenomenal. Let us first say that the restaurant has been in existence since 1887. It is quite elegantly arranged into several eating areas, with lace tablecloths and many old pictures on the walls. The entrance to the eating areas is through the Bar, which has a wonderful historic feel that is accented by an enormous rock fireplace.

Team members began preparing their equipment after an interview with the owner and a friendly exchange of ghost stories with a few of the late night patrons. One of the members of the team had an electromagnetic field detector and immediately detected an exceptional reading near the fireplace. She wondered if the rocks were setting the detector off. Janice's husband, Bill, began taking pictures with his digital camera and Lisa turned on our Panasonic RR-DR60 recorder. Bill immediately found orbs in his pictures, and Lisa later learned that we had recorded an EVP saying, *"Looking to meet forces here. John Coney."*

The walk through of the site took place over two hours, but we captured only about six minutes of recording because of our recorder's voice activation feature. To avoid mistaking the voice of someone in the room for an EVP, we only recorded away from others or when just one person was in a particular area. Otherwise, we left the recorder on "Hold" so that we would not accidentally record.

Employees had seen apparitions around the two wait stations just off of the Kitchen, so Lisa went alone, in the dark with her flashlight, to the closest station and turned on the recorder. With the flashlight, she could see that it was recording something. This area produced an EVP with a male voice that was louder than her voice. It said, *"The Flashlight hunt!"*

The second wait station was very interesting. The energy in that small space felt like a tangible pressure, as if we were being pushed down. A young lady, who told us she was a psychic, was already standing at the entrance of the wait station when we arrived. She did not want to enter the area and only did so because Lisa led the way. On the recording you can hear Lisa say, "Do you feel that pressure?" Immediately a woman's voice said in an EVP, *"Man shot, is sick."*

Later, we noticed a commotion in one of the rooms. Janice had seen a blue streak go by her that could possibly have been an apparition. Several people in the room were taking pictures and reading their meters. Tom was using our Olympus C700 Ultrazoom digital camera, set at an ISO of 800, with the flash turned off. The room was fairly dark with some light coming through the windows from the street. The picture you see here was taken in that room. The bright spots are not light fixtures that were turned on, but reflected light from a glass door. Notice the face in the light on the right. By identifying known objects in the picture, we know that the hand-held camera moved very

Figure 7-1: Available light photograph showing possible phenomenal features

slightly. With this in mind, it is possible that the "latency" of the image in the camera detector may have caused a single point of reflected light to paint the "U" shaped light in the picture. This is a problem one encounters when taking nighttime pictures, as poorly illuminated objects will not expose the film or the digital detector as well as will bright objects. As a result, movement of the camera during the exposure may show up in the resulting picture as a relatively sharp background while points of light may appear to have moved to cause a streak that follows the movement of the camera. The original also shows an interesting electric blue area of light under the "U" shaped streak of light. When the picture is blown up, other ITC features are evident in the optical noise of that area.

After that, we went down a long narrow eating area. *"Oswald James"* was recorded in that area. A girl with another recorder came up to us and we walked through the kitchen together. On Lisa's recorder you can hear the other person's voice ask, "What is your name?" The voice of a man answered back on our recorder, *"Payne and Isaiah."*

In another area we recorded a voice saying, *"That's Niccolites."* Several interesting raps not only preceded this EVP but also occurred after the voice on the recording. Other EVP in this amazing recording session were, *"This is Matt Simmons"* and *"Speak to me. So unhappy."*

Knowing that we would shortly have to leave, Lisa went to another quiet eating area by herself and sat in a chair that was sitting by itself near the wall. She felt that it just looked inviting. Out loud, she said something about how beautiful the place was and that, "I would certainly live here and I know that you are here (meaning the spirits)." Immediately a woman's poignant voice replied in an EVP, *"No, I didn't mean to stay here."*

We followed up our visit to the restaurant with research into the history of the restaurant and the small town of Dayton. A man named John Cooney had lived in the area in 1870, but we have nothing that connects him to this site. The restaurant was first a boarding house that served food, then a bar, and then a restaurant and bar. The area of town around the restaurant experienced two major fires that burned buildings located next door. In the 1930s, a woman shot a man in the back over water rights but he did not die from his wounds. Beyond that, we can not place that happening with this location. Historical information dating back to the 1800s is difficult to find.

On a second trip to the location, we had an additional interesting experience. Concerned about the woman who had recorded, *"No, I didn't mean to stay here,"* we went to the location in the restaurant that the EVP had been recorded and tried to tune in with the entity to offer our assistance. Naturally, we had a recorder going. At one point, we told the woman to look around for a light and to go toward the light where friends would be waiting. After making this statement, an EVP came through in a strong male voice saying, *"We are already in the light."* This of course leads to more questions. This was not the female entity for whom we were concerned. Is it possible that she has stayed in the location because this male has not gone on? Did he tell us that they were already in the light because he wants to discourage any rescue work? Or was this another entity that is able to move back and forth between etheric realms? Only further investigation and recording will hopefully ascertain what is taking place at this location.

Other Investigators' Experiences

Long time ITC researcher, Erland Babcock, told us about his experience of trying to record EVP in a cemetery. Erland and his son made arrangements to go to a cemetery to record for EVP. They had informed the police of their intentions, and as Erland recounts, "That night, when we were trying to make a recording, the police came and checked on us, as a neighbor had reported something strange going on in the cemetery. The police locked up their car and came up to see what we were doing. While they were speaking with us, the lights on the police car suddenly started to blink, the radio went on and off and the overhead lights began flashing. The police just stood there for a while until things quieted down. Needless to say, they walked back to the police station and came for the car the next day. That was the talk around the town for a while. We did not get a single voice and never went back."

=====0=====

Tina Laurent made an appearance as an EVP experimenter in one of a series of shows shown in the United Kingdom called, *The Scream Team.* One of the first utterances she recorded was that of an animal. One of the young adults who appear as a regular in the program, and who is clairvoyant, was bending down stroking a phantom cat. Tina asked her if she had a cat and she replied, "Yes." On the recording between "Cat" and "Yes," a very loud *"Purr"* is heard. The purr was recorded two more times within just a few minutes.

A young medium was also on the program and Tina's EVP voices corroborated what he was saying at the very time that he was speaking the words. At one point the medium was describing a man, whom he was seeing in the corner. The man was telling them all to "Go Away." On Tina's recording a man's voice says, *"Go Away"* moments later.

Tina wrote about another interesting thing happening to one of her recordings during another television appearance, "A few years ago I did a short thing with a local television station in a haunted pub by the seaside. During a break I went over the road by myself to a dark, very old graveyard and made a short tape. I got a loud, clear male saying, *'I killed her.'* I immediately played it for the crew, they all heard it, copied it, yet when I went over the same bit again an hour later the utterance had completely disappeared."

EVP messages disappearing or being changed is reported to us from time to time. It seems to occur sometimes on cassette tapes. At this point in time, it is felt that this will not happen if the paranormal voice is placed on a compact disc. However, we are constantly surprised at the way our spirit friends seem to be able to manipulate technology and so who knows what they will be able to do next.

=====0=====

Susan Bové was raised in a house that had considerable physical phenomena. She wrote that the activity just seemed normal to everyone in the house. This included her parents and three sisters. In 1998, she joined a ghost group and bought a digital recorder. She remembers the first EVP that she recorded, "I was in Mt. Peace Cemetery in Lawnside, New Jersey, and I said into the night air, 'If anyone is here, could you talk to me?' A reply came from a male voice that said, *'Yeah, Miss?'* I was floored! After that, I recorded every investigation and what I have obtained over the years, never ceases to amaze me."

By November of 2002, Susan and six others started their own group. She is the co-director of the South Jersey Paranormal Research (SJPR) group. They have been active in investigating many locations and always have some good stories to share. For instance, the SJPR was conducting an investigation in a Masonic Lodge. Susan was with a group on the second floor when they heard a woman singing near the steps leading to the third floor. A member who was on the third floor also heard the singing and also felt that the sound was coming from the second floor.

Susan radioed other team members and found that no member had been singing. The singing was heard for approximately fifteen seconds. On the tape, you can hear a woman singing, *"I hear you ... are you here?"* You then hear one of the team members ask the spirit if she's female and what she is doing there, as the lodge is a men's club. The spirit responds, *"Singing,"* and then another female voice says, *"She likes to sing."*

The SJPR group is conducting an ongoing investigation in one location that has been found to be very active. It is a private residence in Camden. Susan reports that the group has literally heard spirits and seen apparitions walking around in the house. In one visit to the house, they captured a very interesting event on one of the video cameras! Susan wrote, "The videographer was sitting on the couch, filming into the room. Just to the left, out of the picture's frame, is a Grandmother clock. You can see in the footage, two or more orbs flying out of camera range, towards the clock. All of a sudden, a very loud "BANG" was heard and was captured on the audio portion of the video. The sound was so loud that the camera man jumped! You then see the orbs fly away from the clock.

"Upon inspection, it was found that the weight in the clock was unhooked and had crashed to the bottom of the clock cabinet. The resident's father, who came to inspect it even more closely after the investigation, had made the clock. There was nothing broken. The chain was intact, the hook at the end of the chain was tight and unbent and the hook in the weight was actually a welded hook, so it was all one piece; this was also intact. The only way it could have become unhooked is if someone unhooked it."

Many of the EVP samples that were collected in the house are in a little girl's voice and it is felt they all may be the same little girl. The group had heard a young girl whispering much of the time, but could not hear what she was saying. Susan was sitting on the floor by an open bedroom door. She had just commented on how "pitch black" the hallway had become. After her comment you can hear a young girl say, *"I see now, near to the door."*

On a previous investigation she had been in the same bedroom with another member who had commented that they felt like they had been stepped on, immediately after this a man was recorded saying, *"Sorry."*

The team has also experienced interesting phenomena associated with the family's television set, as well. Once, even though it was turned off, little lights were seen dancing across the screen or running along the bottom of the set. It didn't happen again and the team decided it was some sort of fluke. Then in the last hour of a recent investigation the lights appeared on the television once more. On Susan's recorder you hear the EVP of a little girl saying, *"Did you see it?"* You then hear the homeowner who had just seen it, say, "Oh, My God." Then the little girl again says, *"Did you see it?"*

=====0=====

Karen Camus moved into a house in 1995 that turned out to be haunted. She experienced a wide variety of very dramatic paranormal activities for the three years that she lived there. This experience renewed a passion for the paranormal that she had as a teenager and led to her current research and study in the field.

Karen was with a group investigating a mortuary that was supposedly haunted. The group was sitting around a table preparing for a séance. One of the participating psychics was talking about whether or not people at the table needed to be moved around to achieve a better spacing of male and female energy. When they played back the recording you can hear the psychic speaking but there is also an extra voice speaking at the same time. A strong deep male spirit voice slowly said, *"Poor, I couldn't bury their son."*

This particular site provided quite a bit of material through EVP recording. One particular EVP was in a child's clear voice saying, *"Father."* Quite a bit of information was gathered through the EVP recorded at the mortuary and information obtained by the psychics. With the girl in the mortuary are three other spirits, a male cousin, an adult female and although he has not recorded his voice, the group feels the child's father is also there.

EVP was recorded from the male cousin saying, *"Poor little cousin Bridget,"* *"She still blames him,"* *"Please help her,"* *"Forgive him,"* and *"She'll thank you."* The adult female was recorded saying, *"I'm little Bridget's babysitter,"* and then, *"Go to your daughter."* *"Comfort her,"* *"She needs you."* Karen felt that the last three EVP messages from the adult female were directed to the father and that the child in the recordings may have some unresolved anger toward her father. They are continuing to research this very interesting site.

Sarah Estep has appeared on numerous television shows to record EVP in haunted locations. She was with two London producers and a camera crew in the William Paca House in Historic Annapolis, Maryland. The house was over two hundred and fifty years old and Paca was a signer of the Declaration of Independence. Sarah, of course, had her recorder and was trying to record a few good EVP for a British television program. While they were in the kitchen of the house, with Sarah standing in front of the open fireplace where most of the cooking took place, she asked, "Is William Paca here?" Within three seconds an indignant woman's voice is heard on the recording saying, *"I think not!"*

Unknown to Sarah, one of the producers stood with the curator in the hallway. He whispered to the producer, "Of all the people Sarah could have called on, William Paca is the last. He never went into the kitchen." The woman's voice that Sarah had recorded sounded a little indignant, as if Sarah should have known this.

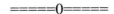

Michele Hardison has been fascinated with the possibility of life after death from a young age. It has been the driving force behind her work. She is the representative in Southern Nevada and Southern Utah for the American Ghost Hunters Society and was the co-founder of Paranormal Investigators of Southern Nevada. Michele has appeared on several television programs and radio shows.

Michele has participated in several investigations at the Pioneer Saloon. It was built in 1856 and is located about fifty miles outside of Las Vegas, Nevada. The Saloon was in the middle of a busy mining camp and was the host to many nightly poker games. In 1915, Joe Armstrong was caught cheating in one such card came and was shot dead. His body lay on the floor for ten hours before the coroner arrived from Las Vegas. The original bullet holes still remain in the saloon.

Michele and her team of investigators recorded an EVP of gun shots being fired. It sounds like something straight from a scene in the "OK Corral." You can clearly hear a loud *"Bang"* then a pause for a split second and then rapid fire, *"Bang, Bang Bang Bang."* A picture was taken the same night in the saloon and showed a paranormal cowboy hat and face. Both of these have been featured on the History and Travel Channels in the United States.

Another interesting piece of history regarding the Pioneer Saloon is a sad visit by the famous American actor, Clark Gabel. Michele wrote, "Carol Lombard's plane had crashed in the mountains behind the saloon. Clark Gabel spent three days in the bar waiting for word on her only to learn that her body could not be recovered."

At an investigation at the Washoe Club in Virginia City, Nevada, Michele and her team recorded the sounds of breaking glass and a woman's voice saying, *"Poison."* The Washoe Club was formerly known as the Millionaires Club. It was known for the wild parties that were hosted upstairs with drinking, dancing and prostitutes. On one occasion, the body of a female who had attended one of these parties was found in the red light district. The death was surrounded by mystery and was unsolved. Was the voice that the team recorded saying, "poison," a clue to the death of this prostitute?

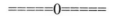

David Vee of Ghosts-UK sent us one of the most interesting EVP recordings that we have ever heard. It was recorded on a MiniDisc recorder in one of England's most haunted castles. On the recording, you hear people speaking in what sounds like Medieval Latin and then you hear the sound of a drawbridge being raised. The site did have a drawbridge but it was removed in 1550.

The MiniDisc recorder was placed in the drawbridge mechanism closet and sealed shut. This closet was the original housing for the drawbridge controls but it is now completely bare. The door to the room that the closet was in was also sealed shut. David and another Ghost-UK researcher, Steve Paton, were the only people on the property at the time the recording was made. They captured this sound eighty five times in a seven-hour period.

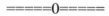

Brazilian researcher Sonia Rinaldi, was contacted by a couple of friends who had moved into a house and then began experiencing terrible problems like sickness, the loss of a car and arguments. They learned that a person had been murdered in the house and were convinced that it was going to be necessary to move.

Sonia asked her spirit team, "I would like to know about the house of C. and M. ... and if you know the person who died there named...?" A male voice came back saying, *"He is already in my house."* Sonia understood this to mean that her spirit friends had removed the murdered man and took him to their house in the Beyond for treatment and rest.

Sonia did not tell her friends of this recording and later wrote to them asking how things were going in their house. They wrote back, "Oh, incredible! It is wonderful! We don't feel bad anymore." Sonia then told them that she had received a paranormal voice telling her that her spirit team had "cleared" the house.

Guidelines for Hauntings Investigation Using EVP

There are numerous ghost hunting techniques to be found on the Internet. There are even correspondence courses that you can take, complete with certifications.

Since our focus here is on EVP, we will not attempt to delineate the accepted procedures for hauntings investigation; but instead, we will describe some of the procedural considerations for collecting EVP in field conditions.

- Rule number one is to always have permission to investigate a location. It may also be necessary to have team members sign release forms to protect the property owner and other team members. Investigations are often done in the dark, in locations that may have hazards.

- Leave everything as you found it. Respect the property that you are on, especially that of historical locations. Know the local ordinances. For instance, it can be illegal to enter a cemetery at night.

- Interview the people involved with the location. Learn about the current situation and what is taking place. Be on the lookout for real phenomena, but also be alert for individuals who like the idea of becoming famous because their house has a ghost. Be discreet and know where to get help if you find yourself in a situation with a family or individual that needs counseling or other assistance that you may not be qualified to give. Know your limitations. Do not try to be an "armchair" psychologist when you are not qualified to do so because you may do much more harm than help. Know the services available in the community and have a referral list with you so that you will be able to refer someone needing help to a professional.

- Take a note pad. It is a good idea to document the experimental session. For instance, note local and sidereal time, local weather, solar and geomagnetic weather, location and any special arrangements made or techniques to try.

- Some investigators tell us that they leave their recorder in one particular location and let it record unattended. Other investigators like to take their recorder with them to the different locations within a site. In our experience, the latter method is preferable. This is because evidence indicates that the experimenter is an important part of the recording circuit; entity, experimenter, recording device. If you move around with the recorder, it will be helpful to verbally record your location each time you move within a site. For instance, saying something like, "I am at the entrance of Building X." Then move on, and again state your new location into the recorder. When you do find an EVP, you will know the exact location it was recorded. This has been helpful for us when we have wanted to go back and try to assist a spirit that seemed to be stuck and had asked for help. Also, knowing our location has often helped to make an EVP message more meaningful and has often provided information that we would have missed or not really understood. EVP recordings may be checked on site but the real scrutiny of the recording will come after leaving the site. Knowing where you were when you collected a voice will allow you to go back to that exact location for follow-up recordings.

◆ It may also be a good idea to record your impressions once you are at the location. Some EVP become more relevant, and can also be a learning experience, when the message can be compared to impressions the experimenter felt at the time. At the haunted restaurant location previously described, Lisa felt a downward pressure in one of the rooms. It is common to feel drops in temperature or to sense spirit entities, but this was a new sensation for her and Lisa's recorded impressions were very helpful. After she mentions it, we recorded a very emotional woman's voice saying that a man had been shot. The next time that we were out in the field and Lisa again felt this sensation, we received a confirmation. In the operating room area at Alcatraz, the downward pressure was again felt and a male voice was recorded telling us that he had died in the room. So do not forget to record any feeling that you may have. Remember that everyone has some latent clairvoyant ability.

◆ Make sure there is a way to distinguish between genuine EVP and the voices of people talking in your area. Operating a second tape recorder or a camcorder, while recording for EVP is a good way to do this. It is not a good idea to depend on your memory, and the ability to compare sound tracks will be useful in eliminating false EVP.

If working with other people, or if there are other people in the area, it is okay to ask them to give you a few minutes of silence so that you can conduct a recording session. If you are not alone, depend on the second recorder to help screen out false EVP.

Additionally, if you are concerned about mistaking a human voice for an EVP, it might help to have a record of the voice of each member of the team doing an investigation and all others that are in attendance. That means a voice recording of the owners and the television crew, as well. If an EVP sounds more human than paranormal it can be compared with the voice record of those present.

♦ It has been found that EVP usually fits the context of the conversations of those present or the questions asked by the investigator. When you record, ask questions of the entities that may be present. When you receive a Class "B" EVP, the context of the question or conversation will help you figure out what is said. Even a very clear EVP will be better understood in conjunction with what is taking place. It has also been proven through experiments that more EVP will be recorded if you ask questions.

♦ Before beginning field experiments, learn about the broadcast radio and television transmitters in the area of the site. It is possible to record programming from nearby transmitters that sound very much like EVP. If you collect relatively long EVP, or if the EVP includes phrases that might be found in transmitted programming and do not fit with the conversations at the site or questions asked, this should be thoroughly checked out before the claim that they are EVP is made. One very honest and dedicated researcher that we know picked up some very interesting and long religious dialog during an investigation. EVP voices usually have a different quality to them than a human voice. This recording was so clear and human sounding that the team checked and found the house was next to a radio antenna. A religious program was being broadcast at the time of the recording.

If a broadcast antenna is found close to a site a solution to possible cross-talk might be to shield the recorder. For instance, setting the recorder to "Record" and placing it in an insulated tin, such as a cookie tin, will help shield it from stray radio and television broadcasts.

♦ Try to work with a team of people that includes someone who is willing to do historical research about the property. You may collect names or information, through EVP or a psychic, which points to a historical event that has already been documented. From personal experience, we know that it is one thing to be an EVP experimenter, but that it is very different to be a skilled historian.

♦ It is useful to have a travel kit containing paper, pen, flashlights, batteries, water and food for snacks, along with the equipment you will use for the investigation. There have been instances in which experimenters have had their recorder and flashlight batteries unexpectedly drained. Bringing fresh batteries to the site, stored in some sort of shielded container, may protect against this. It is not known how the batteries are drained or how to prevent it, but a shielded container is a good start. A coat may be needed, and if working outside, rain gear may be useful. It is a good idea to review the location in good light if you are attempting to record in a very old house or a graveyard—locations that might harbor trip hazards.

♦ Working with a good clairvoyant may offer a way to use cross-correspondence to better understand the EVP. If a clairvoyant was to say, "I sense a woman standing in the corner and she seems sad," and you record a woman's voice saying, "Help me," at about the same time, you would have a very evidential EVP—especially if you have the clairvoyant's voice in the recording. If you behave "as if" you are a medium, and record your location and feelings as you move within a site, we think that you will also be pleasantly surprised with other EVP that become very evidential, as you may find many interactions of which you were unaware.

◆ Ghost hunting is not a dangerous activity. In fact, we have no reports of an entity harming an EVP experimenter in any way. With that said, we do recommend that you consider your attitude during an investigation. Discarnate entities are people too. It is always a good policy to behave as if you are a guest in someone else's home. At the same time, if you tend to be excitable, or if you are fearful of the unknown, you are apt to make yourself a target for a mischievous entity. There is evidence that expectation has an influence on results.

◆ It is not in the best interest of the entity to "hang around" a physical location. Yes, as we have said, it is possible that the entities creating EVP on your recorder are not earthbound or "stuck." But if you feel that an entity may be "stuck," and if the entity asks for help, consider spending time encouraging the entity to move on. A clairvoyant is especially useful in such rescue work.

◆ If using a camera or video recorder, be aware of the possibility that you may be stirring up considerable dust, and that airborne dust can cause orbs in your film or recording media that are easily mistaken as evidence of an entity. It does appear that some orbs are evidential, but it is very difficult to distinguish artifacts of this sort from true evidence. Our recommendation is that the camera should be used with available light if possible, and that the camera flash should be used as little as possible to avoid such artifacts. Remember to check the audio of the video recording for EVP.

◆ Show respect for the living. Many haunting clubs make it a policy not to attempt an investigation in a location of a recent tragedy. Be aware that the loss of loved ones will be fresh and attempts to contact the deceased may not be well received. When a tragedy takes place, such as what occurred on September 11, many experimenters at distant locations have picked up references to these events through EVP. It is not necessary to go to the location of the event. If you want information about such a tragedy, simply try recording and asking questions about the event wherever you are.

♦ It is a good practice to be sensitive to the feelings of others, should you collect an EVP that seems to pertain to a missing person or some other situation that might invoke strong feelings in a loved one. First, do not give people a message via EVP unless they have asked for the message. Next, if you are not one hundred percent confident that your EVP is accurate, respect those who are grieving and do not attempt to publicize the message. Be mindful that some entities will tell you anything you want to hear, especially an individual that is looking for publicity and has no ethics.

♦ People who invite you into their home or onto their property for EVP experiments are often curious about what you are doing and whether or not evidence has been collected. They will probably want to know your conclusions as well. It is always a good idea to offer a follow-up report, and it is very important that if you agree to such a report, you follow through. When you do make the report, tell them what you know. If you are speculating about one aspect of the report or another, tell them that as well. It is better to admit that you "think" something is true, rather than to leave them with the false impression that you have reason to know something is true.

Earthbound Entities

Some EVP are initiated by local entities such as those that have stayed in the physical locale for various reasons and have not crossed to the other side. A friend and medium, Vickie Gay, recently began offering readings at psychic fairs. She had not thought that she would be conducting rescue work but has already found herself in this situation. She commented that one entity, that came with a lady for a reading, had not moved on because he was afraid that if he did, he would not be able to see his family again. Vicky worked with the young man and convinced him that, if he went on, he would be able to progress and grow and still be able to contact his family through various methods, such as through a medium or through EVP. In reality he was earthbound and was following family members around, but they did not know this. Vicky was able to convince the entity that, in going on to the other side, he would be able to find others interested in contacting those still on the physical side of the veil and be able to learn from them.

Another funny, yet sad, entity that Vickie also found herself helping was one who did not feel that he could move on until he saw everyone floating up to the sky in the Rapture. His belief that Christ would appear in the sky above the earth, and in a flash, every "saved" believer would disappear from earth and be lifted into the sky, had prevented this person from moving on to his true state of being.

You can see from the above two situations that every entity represents a unique situation, but we have found that there are a number of typical reasons why entities may become stuck.

- They may not know that they are dead. Many EVP recorded at haunted sites are those of children and we feel that many of these did not understand death.

- Alternatively, they may realize that they are dead but are afraid to cross over because of the possibility of judgment, and the possibility of being cast into hell, for some act that they have committed in life.

- Their religious beliefs are such that they feel they can not move on. They are waiting for a certain event to take place on earth, as this is the time when all people of their religion will cross to the other side together.

- Based on EVP messages we have collected, some entities have remained near the physical locale because another discarnate entity that they are close to refuses to move on. This can happen with groups of people, such as families or soldiers. If one is conducting rescue work in a situation where there is a group, the entity refusing to go must be the one worked with. If that entity is helped and convinced to move on, all others in the group will usually follow.

- Stuck entities are often associated with a traumatic or emotional event, such as a violent death or an unexpected transition that has left them with a sense of "unfinished business."

- You may also find entities who have remained in a locale because of a perceived injustice that might hold them until it is resolved or until they have been convinced that it will not be re-solved.

♦ Some entities simply do not wish to abandon earthly possessions or situations which they loved in life. This latter group may simply crave physical experiences or sensations or be deeply attached to their home or some other possession.

Not all Ghosts are Entities

Not all ghosts are self-aware entities. Look back at the EVP example described on Page 136 in which the sound of gunfire can be heard, and the sample on Page 137 in which people speaking and the sound of a drawbridge being raised can be heard. These are examples of EVP that are clearly initiated by what is commonly referred to as a "recorder ghost." Traumatic events are thought to cause a record of the event to be etched into the fabric of the location. It is not clear what triggers these memories to "replay" the event, but the result is that witnesses sometimes see ghosts that are not responsive, and that repeat the same activity time after time. These reenactments are sometimes caught in recordings.

In this case, the person who speaks in an EVP message, or the person who is seen as an apparition, is thought to be little more than the manifestation of residual energy left there by an emotion-filled event.

There is a synergy

Field recording for EVP can be a very enjoyable pastime. EVP is a potent tool for hauntings investigations, and when used in conjunction with the other tools in the investigator's arsenal, it offers interesting information and an effective way to better understand the nature of the haunting activity.

EVP and ITC research share a common goal. Both are concerned with discovering the nature of the greater reality. Both are important tools in our search for proof that our personality survives bodily death. Investigation of hauntings is considered by some to provide an important introduction to the world of paranormal phenomena for people who might otherwise never consider that there is anything more to the world than meets the eye. We, who study EVP and ITC, find hauntings investigations and ghost hunting not only fascinating but also a valuable source of voice and image phenomena that prove there are many things going on in the etheric space around us that we can not physically see or hear.

Chapter 8

Results from Current ITC Researchers

Erland Babcock - Video ITC Images - United States

Erland Babcock has been conducting research in EVP and ITC for many years. He was one of the members of George Meek's research team. Erland has experimented in capturing images and voices in many different ways, thus it is difficult to tell the reader of an exact equipment setup that he uses.

Erland has told us that he operated a television recording studio in which he used a Panasonic Vidicon camera. The camera was focused on a twenty-five inch commercial, high-resolution monitor/receiver in the monitor mode. The output of the camera was fed through a Time Base Corrector, which enables him to "freeze" a video frame, then through a video enhancer and then back to the monitor. Thus, there was a loop in which the camera "saw" what the monitor displayed and the monitor displayed what the camera saw. Erland normally operated the camera just a few inches from the picture tube and slightly out of focus.

Erland normally recorded for just a few minutes and then reviewed the tape, a frame at a time. He took a photograph of the most interesting frames with an ordinary camera. Erland did not specify whom he wanted to see nor did he know any of the people or the scenes in the pictures. Not all frames had features, and sometimes a week would pass without a feature. Some features appear to be bird's eye views of clouds or land. You can see examples of his work using this technique at aaevp.com.

Erland has also collected features by simply photographing noise on a blank

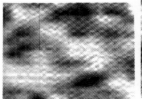

Figure 8-1: Video ITC features collected by Erland Babcock by taking a picture of television static.

television channel with a digital camera. He wrote, "I was using a blank television channel on my computer. I used my digital camera to photograph the moving snow on the screen. I did this several times at random times. I then removed the television channel from the computer screen and loaded the images from the camera into the computer." Erland brings up each image one at a time and uses software to remove the color so that he has a black and white image. He then searches the image for anything that looks like it might be interesting. When he finds something, he crops it to remove the rest of the frame, and enlarges the feature one to eight hundred percent. He wrote, "If the feature remained interesting, I enhanced it by the use of contrast, brightness, sharpness and any other control that will make it more recognizable.

"These pictures (Figure 8-1) were produced that way and are from nothing more than random noise on a blank television channel."

Marcello Bacci - Radio ITC Voices - Italy

Marcello Bacci, of Grosseto, Italy, became interested in the paranormal in 1949. Soon after this, he began recording voices using an old vacuum tube radio tuned between seven and nine megahertz as a sound source. Marcello has been contributing to research in ITC for over thirty-five years.

Word of his early results spread and people frequently stopped by his lab at home, and often had their departed loved ones talk to them through Marcello's radio. These early sessions soon developed into Friday evening group meetings attended by many people. The Marcello voices are often so loud and clear that everyone in the room can hear and interpret what is being said. Contact with the other side often lasts for over an hour. Questions are asked by the sitters and some of the questions are immediately answered and can be heard clearly over the loudspeaker. Other answers from the entities can only be heard by replaying the tape, which is used for documentation.

Marcello does not only receive paranormal voices. The spirit group ends each session with singing. It may last for over a minute and sounds like a heavenly chorus.

In one of the earlier experiments, Marcello and his colleagues took two identical receivers and placed them about a meter apart and achieved contact with the invisibles. Both were tuned to the same frequency but only one of them brought the paranormal voices in. They changed the tuning on that radio while reception of the voice was being heard and the voices continued to communicate.

Professor Mario Salvatore Festa writes that he has been privileged to witness Marcello's experiments for six years.[75] He explained that the session begins when Marcello calls for the door to be closed and the lights turned down. Everyone waits while he turns the knob on the radio looking for a zone of white noise. Finally, he announces that he hears them. Immediately something sensational happens. All radio signals stop and a sound like the noise of the wind is heard. Marcello talks to the radio and calls out, "Friends, we are here, can you please make yourselves heard?" And then, the voices are heard. They speak clearly and answer questions. Many messages are filled with esoteric content. Children who have crossed over speak to their parents and the room is filled with extraordinary emotion and joy.

During experiments in April 2002, Mario brought professional tools to measure the electromagnetic field to gain a better understanding of how the phenomena come about. Measurements were taken with the radio switched off and with it on. In the instant when the voice phenomena started, there was no significant variation in the electromagnetic field.

Festa writes, "We began what was to be a surprising and shocking experiment to the astonishment of the people present, especially the mothers ... Electro-technician Mr. Franco Santi ... with my agreement, took out two valves [vacuum tubes] from the radio while the experiment was taking place. Firstly, he took out the valve that controls the frequency modulation .... This did not have any impact on the working of the radio, as the receiver was tuned on short waves.

"Then, as the people present continued to look and comment with disbelief, he took out the second valve, a local medium frequency oscillation converter. This silenced the short waves completely.

"It was then ... that I could move the tuning knob up and down the frequencies to find 'the absence of a signal' while the connection with the voice from the other dimensions continued unchanged.... What was happening? The rule of standard physics had been turned upside down, the 'entities' continued talking as if nothing had happened."

Diana and Alan Bennett - ITC Pictures of Other Dimensions - United Kingdom

Most people who are interested in the paranormal have heard of the Scole Experiments. The four members of the Scole Experimental Group (Sandra and Robin Foy and Diana and Alan Bennett) sat over a period of four years. The spirit team at Scole was prolific, using a new Spirit World technology that did not require the use of ectoplasm for the production of physical phenomena. The use of ectoplasm can be dangerous for the medium, while this new energy was a blend of Spirit World energy with earth energy and did not involve ectoplasm at all.

Phenomena experienced by the Scole group included the appearances of solid spirit beings and fifty small objects received as apports. People, such as psychic researchers from the SPR, witnessed physical phenomena including pictures, handwriting, symbols and messages that appeared on factory-sealed photographic film.

The phenomenal success of the group and their spirit team resulted in the book, *The Scole Experiment*,[66] by Grant and Jane Solomon. The group stopped sitting together for experiments in 1998.

The Scole mediums, Diana and Alan Bennett, have spent the time since the last Scole session exploring their own particular interests in psychic phenomena and healing.[76] They felt that they owed so much to those in the higher realms and knew in their hearts that they would continue to experiment in one way or another. Alan wrote, "I have always seen in my mind's eye, a phoenix rising from the ashes with renewed vigor

Figure 8-2: Woman—The Bennetts have been given the name of this helper on other side

to live through another cycle. Diana has always had that wonderful gift of 'far sight' and receiving guidance in her dreams, so it was no surprise when she told me that she had 'seen' the two of us working together again. She was told how we should take the first step in the form of a shared experience. We followed her vision of a simple experiment using a crystal, where the two of us linked together mentally and shared a journey of exploration.

"This experience was incredible; as we were both shown such wonders and places. We were guided by a sentient being through a wonderful 'magical mystery tour' where we met another being whom we conversed with. Through several other forms of communication, we were shown how to set up an experiment that would, they explained, enable us to glimpse into other worlds (dimensions). We found that it was necessary to continue to follow our intuition, and to be 'guided' by them, if we were to achieve satisfactory results. It has required total dedication and perseverance as well as an open but still discerning mind.

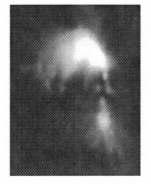

Figure 8-3: Old Man collected by the Bennetts

"Before I continue, I would like to mention one very important factor, and that is that old chestnut 'total darkness.' We were 'guided' to do this new work in full light. After so much criticism in the past regarding physical phenomena obtained in total darkness, we decided that we would only continue to work as mediums in full light. This has proven not to be prohibitive in anyway whatsoever, as we have achieved very encouraging results.

"These experiments are primarily attempts to see into and capture visions of other dimensions of existence. To be more precise, these visions are more like frozen images or pictures of different moments in time as we look into these dimensions."

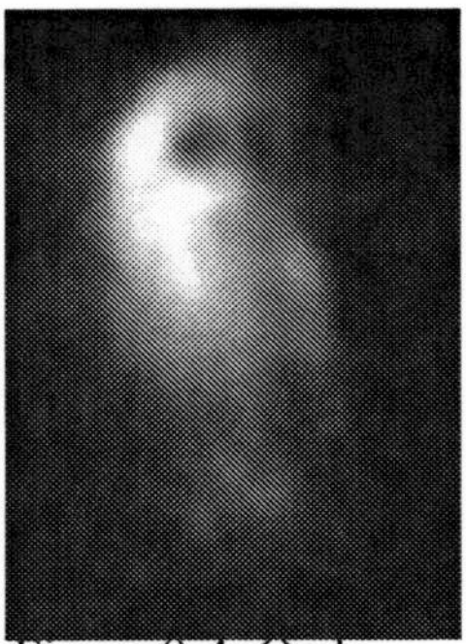

Figure 8-4: Owl collected by the Bennetts

Since that first experiment, the Bennetts have conducted many more experiments, progressing and building on what they have learned. Alan told us, "It seems that the possibilities are endless, as we modify and introduce other equipment into the experiments. The experiments are based on the idea that there are dimensions not only beyond our own but within them also. By using a combination of electronic and photographic equipment coupled to image enhancement computer software (for magnification purposes only), we have been fortunate enough to obtain fleeting glimpses into these 'other' dimensions during our experimental sessions. The experiments also require focused visualization by the two of us to create a central focal point for the experiment."

Diana told us that, "The images pictured here were obtained with the use of an ordinary camera and a good lens (not digital). The process requires us to strategically place crystals, according to instructions given by Spirit. (This varies with each experiment.) Light is focused over the crystal area, some through mirrors, colored filters and/or reflective surfaces. We intuitively know when it is the optimum time to take the picture. When the pictures are developed, they are digitized, enlarged and then examined using Photoshop.[22] As you can see, (in Figures 8-2, 3 and 4) the images are fascinating. Sometimes, the images take up most of the frame sometimes only a small part."

What excites Alan is that, "This new approach is only now possible with the advancement in electronics and the related computer technologies. Therefore, where will we be and what will we discover in the years ahead?"

Paulo Cabral and Phyllis Delduque - Video ITC - Brazil

Phyllis Delduque and Paulo Cabral are two Brazilian researchers who have been working with Instrumental Transcommunication since 1998. They work with both voices and images. The researchers say that ITC not only changes one's ideas about the continuation of life but also transforms the consciousness of the individual. Their contacts from the other side share teachings as well as messages from friends and relatives who have already gone to the other side of life. Phyllis and Paulo feel that the majority of what they receive through ITC is related to spiritual mediumship, with the other factors being the equipment that they use and their applied methodology.

Phyllis and Paulo use a video camera that is connected to a video recorder. The output of the video camera is connected to a television. The camera is focused on the television. This arrangement provides a record of the process while supplying the necessary feedback for the video camera. In effect the camera is "looking" at what it looked at a few milliseconds past.

Anabela Cardoso - Radio ITC - Portugal

On March 11, 1998, Anabela Cardoso was home alone with her dogs. She tried making contact via EVP at around 7:00 p.m. This was the usual time that she had been trying to make contact. Suddenly, from one of the old valve radios tuned in to white noise, a loud voice shouted: *"We are listening to everything! We want to know about the world, we want to hear your things! Now we are going to count on you to offer what is fair! I was not the one who spoke, but I suppose that you have made a question! This is very and very difficult! Another world!"*

When the voice came, Anabela had just asked if the contacts that they were getting through EVP so frequently were really coming from Timestream Station (some messages said they were) and from Carlos de Almeida. She knew that Carlos de Almeida was active at Timestream and she usually asked for his help and protection at the beginning of each recording session, as she still does.

From that date on, the voices came frequently from one of the radios tuned into the white noise of short waves. On a few occasions the voices would start on one of the radios and then suddenly "jump" to another radio tuned to a completely different frequency.

Carlos de Almeida and the voices of Timestream speak about a wonderful world very similar to ours. They say that they have new, young and vigorous bodies and that they can travel by way of their thoughts. They have told her that they modulate radio waves with thought and that this takes a lot of effort and concentration. Anabela was told that it is their thoughts that become audible and that their thoughts come, in a way, like radio waves.

Anabela writes that, "This is probably the reason why there are so many semantic and grammatical mistakes in direct radio voices, why there are words of different languages mixed in the same sentence and why they jump without stopping from one subject to another in the conversation, they can speak among themselves without stopping for hours, and why that often an answer to a question asked from our side becomes glued to their own conversation. Maybe this is how thought sounds."

The confirmation of Carlos de Almeida's identity had come several times in answer to her repeated questions, and she also heard from close relatives who identified themselves by giving their names. It was announced that Timestream Station would begin transmitting regularly to her on June 1, 1998.

Most of the messages that come to Anabela, using what she calls the Direct Radio Voice, or DRV method, are in Portuguese and so the examples given in the following text have been translated into English. Some of the information that she has received from this group is:

- There are no conditions at all attached to contacts. A normal person who is sufficiently interested and persistent can aim to communicate (with the other world).

- The most important thing to establish contacts is *"Contact itself."*

- The level of transpartners on the other side always corresponds to the ethical and intellectual conditions of the earth partners.

- Meditation is very important.

- They have a physical body of a different nature. They can feel the sensations perceived by our physical senses.

Some of the information that they have given Anabela about what their world is like is, *"It is beautiful here,"* and, *"It is the gate of light."*

Anabela's contacts have told her that they can be in their world and in ours at the same time. They also say that they have contact with other worlds besides the earth.

The entities urge ethical treatment of animals and also plants saying, *"Don't forget that the plants are beings of your world. From that world all come to this world."* In answer to a question regarding the ideals and goals of animals in their world they said, *"They also try to know more."*

Larry Dean and Patricia Begley - Video ITC - United Kingdom

Author Larry Dean has been working on a book on Instrumental Transcommunication. Larry told us that he was interested in spiritual pursuits at a very young age when others were out playing sports. Later in his life he devoted most of his time to making presentations and teaching classes on the development of psychic abilities and healing. Although Larry is naturally clairaudient and clairvoyant, it was only after he stopped teaching these classes that he took time to develop his own trance work.

  Larry is a deep trance channel. His ITC partner, Patricia Begley, facilitates and records the sessions. It is generally only through her doing this, and the recordings she makes, that Larry knows what has taken

Figure 8-5: Patricia's Aunt Ethel

Figure 8-6: Video ITC of Ethel collected by Larry Dean and Patricia Begley

place while he has been in trance. Many years ago, the two were told by their main guide, Choi, that their spirit team was working with them to establish Video ITC contacts.

Larry and Patricia were excited and encouraged by the ITC results of the Harsch-Fischbachs, and were even able to sit in on some of the Scole Experimental Group sessions. They continued to sit with a group and did experience limited physical phenomena, but did not achieve successful television or Video ITC contacts. Larry talks about the years that they sat around looking at dots on television screens and featureless video experiments in an effort to find phenomenal features.

Larry first made contact with us to ask for information about ITC for a book he was writing. Through subsequent telephone conversations with him, we have found a common bond of interest in ITC. Since we had also begun work on this book in late 2002, we were able to send Larry the two chapters on Video ITC that describe how to conduct an experiment. Just weeks after he received the instructions, we received an excited phone call from Larry. He and Patricia had followed the instructions and had received images on their very first experiment. One of those pictures was that of a close friend who is now on the other side.

Larry has the wonderful benefit of being able to use his trance work to receive advice and direction on future experiments, and through his trance work, they were told to ask for particular people to appear in their experiments. Even though they have only been working with Video ITC since late in 2002, they have succeeded in receiving many very good features, including images of people whom they have asked to appear in their video frames. For instance, they have collected images of Patricia's father and mother and also the image of Patricia's Aunt Ethel, which is shown here.

Luis de la Fuente and Estrella Fernández - Video ITC - Spain

Figure 8-7: Gordita, collected by Fuente and Fernández

Luis de la Fuente and Estrella Fernández, of Madrid, Spain, have been conducting research in ITC since December 1989. They use a video feed-back loop connecting the television (without antenna), video camera and a VHS video recorder into a circuit. Luis writes, "We see the television screen when looking at the viewfinder of the video-camera. At this point, and by means of the video camera zoom, we obtain electromagnetic oscillatory 'clouds.' We then record on the video for one minute. We rewind the video and study the tape frame by frame." The two call this series of four frames, "Gordita."

In other experiments the two have turned the video camera to a forty-five or ninety degree angle, or they face it downwards and record through a mirror. Luis and Estrella have experimented with using filters of different color in front of the camera. They have recorded in broad daylight, with a tungsten light, infrared bulb, black light, and in total darkness. They always use the video feedback method with all of these. Their television is an old valve [vacuum tube] model.

Luis wrote, "In short, we don't know how it is possible to obtain a moving transimage because "Gordita" and a few others are rare exceptions. Change? An effort from the other world?"

Pascal Jouini - Video ITC - France

French researcher, Pascal Jouini, experimented first with EVP and obtained results that proved to him that the phenomena were real. He did, however, stop his experimentation in EVP because the sessions took a great deal of his energy and made him tired. On occasion, we have heard this from other experimenters. One experimenter noted that she really noticed EVP taking her energy away when she was suffering through a long illness. She found that she had to stop recording during that time. Knowing that the experimenter is part of the experimental circuit, EVP experiments should be conducted when your energy is high and you are in a positive frame of mind. In our work with these phenomena, we have not noticed a drain in energy and actually feel that we gain energy from experiments.

Some months after Pascal's first EVP experiments, he began experimenting with Video ITC using the black and white Klaus Schreiber technique. He did not, at first, find anything of interest in the video frames and began trying variations to the Schreiber method. He found that he had good results when using color, a speedy flashing on the screen with the camera tilted sideways at a ninety degree angle to the screen. This gives Pascal an "X" pattern on the screen due to the difference in the angle of the scanning trace between the television and the camera. He feels that he is getting more contrast and more faces with this method. He uses Photoshop Lite[22] to analyze the pictures and to find hidden faces using the light and contrast functions.

Pascal has also successfully experimented with a video feedback loop using his computer monitor. This removes the television from the circuit and proves that the ITC features are not caused by television programming.

As this book is being written, Pascal sometimes runs an online webcam for video ITC on his website. This is commendable because he is able to demonstrate a live Video ITC experiment for people anywhere in the world. Visitors are not only able to watch but also can capture their own ITC features. He sometimes has the webcam operating on Sundays and we have been able to access the experiments at 11a.m. Pacific Time in the United States.

Pascal's website includes many examples of his work, along with the live web cam. There are also examples of the work of other researchers. The address is:

http://perso.club-internet.fr/pjouini/menugb.htm)

Mark Macy - Luminator Device - United States

Mark Macy's work can be described as a form of Photographic ITC. Very interesting features may be found by photographing people who are standing in the field of energy generated by a subtle energy device called a "Luminator." As Mark described the Luminator, "It is a tower-shaped device about four feet high. According to its inventor, Patrick Richards of Battle Creek, Michigan, it alters the environment in an area that extends about one hundred feet in all directions from the device. It has two internal fans, which move air over a series of liquid-filled rings. As the air molecules go tumbling through the rings, unusual things happen to them, according to Richards, including a reverse spin of electrons.

"There are eight Luminators in existence today which are being used in several countries. All but mine are being used for healing, as the device seems to have some important therapeutic applications. Mine is the only one used exclusively for ITC research, as it provides a very simple and basic means of ITC contact. When I take a picture of someone with a Polaroid camera in the presence of the Luminator, there is a good chance that there will be other faces in the picture

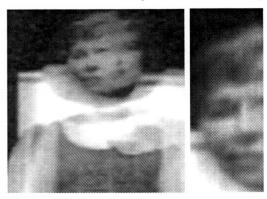

Figure 8-8: Polaroid picture of "Joyce" (Left) taken by Mark Macy in the Luminator field. At the right, just the right side of Joyce's face showing a resemblance to the Singer, John Denver.

after it develops a few minutes later. These are faces of people who are not physically present; spirit beings.

"I like the fact that the Luminator seems to melt away some of the subtle barriers between our physical world and the spiritual worlds that are superimposed over our reality. That superimposition of realities is a belief of many religions and a theory of many modern scientists and researchers, such as the late George W. Meek. Mainstream science disregards these theories and beliefs, for the most part, but I believe that this disregard will become harder to justify in the future as evidence such as the Luminator images come to light."

Mark had been corresponding with friends of John Denver, the well-known singer who died in a plane crash on October 12, 1997, in an attempt to locate pictures of the late singer. One phenomenal image that Mark received at a conference in Colorado Springs in April, 2002, bears a strong resemblance to the late singer. Both Jack Stucki and Mark had Luminators running, and took pictures of about 30 workshop participants. Many of the Polaroid pictures had spirit faces posing with the human faces. Figure 8-8 is a picture of "Joyce" taken in the Luminator field and a blow-up of a spirit face that resembles John Denver. You can see these pictures in color at Mark and Rolf-Dietmar Ehrhard's excellent website at:

www.worlditc.org/d_02_more_proof.htm.

Mark explains why he likes the technique he is using, "One tremendous asset of this spirit photographic process is that it seems to work best with a Polaroid camera. This eliminates the chance of a hoax, since the photographer, the subject and spectators are all present as the picture is snapped, as it develops, and as it becomes clear a few minutes later. Many people today are enmeshed in the physical world and have no knowledge of or belief in spiritual reality. My hope is that these Luminator images and other good results of ITC research will help people reconstruct their mental roadmaps or models of reality in order to accommodate a healthy spiritual view."

Jutta Liebmann–Photographic ITC–Germany

Jutta Liebmann, board member of the EVP and ITC group in Germany known as the VTF,[53] sent us some of her pictures of extras. "Extra" is a term that describes the presence of entities or objects in photographs that were not there in the physical sense. Jutta wrote, "I have taken these photos from my television screen with my Olympus Standard Camera AF-10 Super, using the flash attachment. During most of the experiments I did not notice anything paranormal."

Jutta sent us several excellent pictures. In Figure 8-9, Jutta had videotaped a documentary about the early beginnings of humankind. At the end of the program an earth view with clouds was shown. "When I played back the videotape, I discovered this paranormal

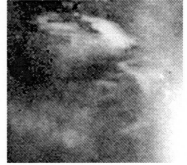

Figure 8-9: Photographic "extra" taken from a television program by Jutta Liebmann

face in the clouds. I then put the video recorder into the stop/standby position and took the picture."

The features in Jutta's photos look a lot like the features experimenters find with the video feedback method. Video ITC also shows extras, even though Video ITC is accomplished with a chaotic video signal, rather than photographing a physical or video scene. It is fascinating to see Video ITC-like features in photographs of a taped television program. This gives us good reason to look closer at other photographs and video recordings that include visually "noisy" scenes.

Sonia Rinaldi - Telephone ITC - Brazil

In 2001, Sonia Rinaldi[35] began helping parents who have had children transition to the other side by making phone calls to the beyond. When asked how she informed the Beyond that a certain parent would call on a certain day, and that their child would be needed on the phone line, Sonia responded that she did not know, but explained that, "What seems to happen is that the 'Beyond' is here all the time. It seems that they control everything around here. This makes me imagine that, in the parallel space of my house, many spirits live and work in the station. [Sonia is referring to a spirit side transmitting station for ITC] Maybe the station is over my house. I don't know. The fact is that, when a person is put on the agenda, that is all that I do … on that specific day, the expected deceased appears and replies with particularities that only the parents know, and also, in the great majority the voice is recognized."

Sonia has tried to make phone calls while in other countries and has also tried having people from outside of Brazil call in for a session. This has not worked. She wrote, "For me this means that each country or big region has its own leaders on the other side. Its own stations ... I say this because maybe an AA-EVP member may wish to make a phone call with me. It will not function if he or she is not from Brazil.

"Not only that. The person apparently must be Brazilian, or the deceased must be Brazilian. On September 11, when I saw on television that many Americans were dying, I made a contact by phone and asked, "Are you (in the station) receiving many people today? Are you receiving Americans?" The reply was, *"This is a station of Brazilians."*

Sonia had made over one hundred and sixty nine well-documented calls as of July of 2002. The Noetics Institute Incorporated (NII)[64] in the United States is now supporting Sonia's research. Here is a typical transcript of one of these calls.

Arranging a Phone Call

The parents make an appointment with Sonia to place the paranormal phone call in advance. They are instructed to prepare ten questions. When they call on the appointed day, Sonia has one phone in her hand and leaves an extension phone open so that those in the Spirit World may participate. The telephone is connected into the microphone jack of her computer and all conversations are recorded directly into the computer. The questions are asked leaving ten seconds between each question and the recorded conversation lasts about twelve to fifteen minutes. After the session is over, Sonia prepares a final recording of the session resulting in seven to eight minutes of pure dialog between parent and deceased child. The sound track is copied and sent to the parents who are then able to hear their loved one's voice and decide if it is their child who is speaking. The parents often identify not only the voice itself, but also details and information which only they knew. Sonia also requests that the parents send back a report detailing their impressions of the recording.

Here are excerpts from a typical transcript of one of Sonia's telephone calls to the Beyond.

Recording Number 18—Historical Background

Luiz and Virginia, both medical doctors, lost their two only daughters, Luciane and Viviane, in an automobile accident in November of 2001. This is exactly one year from the date that this call to the other side was made. Two others who died in the same accident were the cousin, Carolina, and the Uncle Fernando who was also a medical doctor. Fernando was Virginia's brother.

There are two different female voices that are recorded during the call. One is soft and sweet and the other is more energetic. Also a large number of the voices on this particular call answer before the questions are asked.

Sonia conducts a test of the equipment before the parent's phone call is received. Her main contact on the other side is a gentleman who speaks Portuguese with a German accent. Sonia simply calls him "Mr. German." In the following excerpts from a transcript that has been translated for us by Sonia, all paranormal voices are in Italics.

Test before the Phone Call: Sonia takes the picture of the two girls and raises it in her hand. "Hi Mr. German, these are the girls."
Male voice: *"I can confirm!"*
Sonia: "Are you ..."
Male voice: *"I can hear you."*
Sonia: "... hearing me well?"
Mr. German: "They are already here!"
Sonia: "I would like to know if Luciane and Viviane ..."
Male voice: *"They are in contact."*

...

An interesting thing happens here. Sonia was contacted by the wife, Virginia, who made the appointment. Sonia thought that Virginia would be the one making the phone call and is surprised when the call comes in from the husband instead. The following was recorded before he called and shows that the girls knew they would be speaking with their father.
Young girl: *"He my father!"*
Sonia: "Hi girls ..."
Young girl: *"Hello!!!"*
Sonia: "Hi Luciane ..."
Young girl: *"We are here—in the challenges using ours day by day."*

The Transcript of the Phone Call: Sonia: "Good morning friends, we are with Virginia and Luiz on the line, anxious to talk with Luciane and Viviane ... and we trust that they will answer and will bring happiness to the parents...."
Another interesting thing happens. Sonia would usually say to the parents, "Okay, you may go." This time, however, she says something different and to her surprise the young voice says what she normally would say.
Young girl: *"You may go!"*
Sonia: "Ok Luiz, you may begin!"
Luiz: "How are you there in the Spiritual Plane, daughters?"

Young girl: *"It is the proper peace."*
Young girl: *"I will call you up!"*
Luiz: "Are you both always together?"
Here in the dialog, one of the most beautiful exchanges occurred as a soft voice spoke with feeling.
Young girl: *"You called us here and I heard you!"*
The delicate tone reveals that she had heard the parent's suffering as a force that called her to them.
Luiz: "Little daughter, Uncle Nanão and Nina ..."
Young girl: *"Yes they are!"*
Luiz: "... Are they well?!"
Young girl: *"Uncle Fernando was informed!"* (Informed of the call.)
Young girl: *"This date was a vision!"*
Young girl: *"I came to enjoy it!"*
Luiz: "Daughters, the father and mother loves you a lot. Is there anything we can do to help you?"
Young girl: *"I am the responsible for my own peace!"*
Luiz: "Have you seen grandpa Ângelo?"
Young girl: *"He is beside you."*
Young girl: *"He already arrived!"*
Luiz: "Who is with you at this moment? Is uncle Nanão there? Does he want to speak anything for us?"
Young girl: *"There exist people beside you."*
Young girl: *"He is prepared."*
In fact, a male voice would speak close to the end of the call—possibly uncle Nanão.

...

Luiz: "How is it the routine there, little daughter? Do you usually eat? Do you sleep?"
An older female voice enters to command and to inform about the girl's limits.
Older Female: *"They can just speak!"*
Luiz: "In what activity you are in charge of now, little daughter?"
Young girl: *"Something very crazy!"*
Young girl: *"That is correct, my father!"*
Luiz: "Do you like your activities up there?"
Young girl: *"Take care!"*

Young girl: *"I missed you for your love and affection!"*
Older Female voice replies: *"Yes, they do!"*
Luiz: "Do you already have a lot of friends there?"
Older Female: *"She is getting tired!"*
Sonia notes that this is not the first time that they have revealed to her that the contacts are tiresome and she wonders if it is the proximity to the earth that drains their energies.
Young girl: *"Lu kissed you in fact!!"*
The father called Luciane, "Lu."
Young girl: *"Unhappily not!"*
Luiz: "Do you leave for walking frequently?"
Young girl: *"Only to see you cry!"*
Luiz: "Do you have permission to visit us?"
Young girl: *"Yes in the everyday!"*

...

Luiz: "... receive the love of your father and your mother! A big kiss!"
Luiz: "We will always be together forever...."
Young girl: *"We count on that."*
Luiz: "God allows us!"
Young girl: *"For sure, my father!"*
Sonia: "Viviane, can you say where you are at this moment?"
Older female voice informs: "She is leaving!!! She is totally weak!"

...

Sonia then tries to speak with the girl's uncle, who died in the same accident. Once again the answer comes before the question.
Masculine voice: *"I am already speaking here! It is clear!"*
Sonia: "Would it be possible for us to speak to Fernando?"
Virginia: "Fernando ... leave a message to our father and mother"
Masculine voice: *"I will! In spirit, like a new man!"*
Virginia: "Are you happy where you are?"
Masculine voice: *"Friend! It is easy!"*
Sonia: "OK Fernando ..."
Masculine voice: *"Oh, I was so happy recording!"*
Sonia: "Our great gratitude. We thank mainly Viviane ..."
Masculine voice: *"Hello is here!"*

Sonia: "... and Luciane. We thank all the friends of the Station and we are interrupting this recording."

Sonia asks each parent of a phone contact to fill out a questionnaire and send it back to her. When doing this the parents expressed their deep gratitude. They say that they noticed the difference between the two young girl's voices and that they coincide perfectly with the characteristic of their two daughters. They wrote, "Luciane, without losing her sweetness and, in spite of very calm, was always more emphatic and fast when speaking. Viviane was always more slow and sweet and calm when speaking."

The parents also say that Fernando's expressions and voice were true to his life time personality. They told Sonia that others in the family had heard the voices and agreed.

These phone calls are made in Portuguese and are translated into English by Sonia. This is only one example of several fascinating transcripts that we have had the pleasure to read. Each call has been unique and provides us, not only with evidential information, but also with a glimpse into life on the other side.

The Noetics Institute Incorporated (NII)[64] has arranged to sponsor Sonia's work during a study period in which Sonia will follow an experimental protocol designed by the NII scientists.

Chapter 9

Our Experience with Video ITC and Other Phenomena

The paranormal pictures collected by Klaus Schreiber, Adolf Homes and the Harsch-Fischbachs have always interested us. In fact, we had purchased a video capture card for the computer while we were in the fifth-wheel, but we never got around to trying it with our camcorder. It was difficult to work with a video feedback loop in our home on wheels, as the television was built into an overhead cabinet. Looking back, it seems clear that we were receiving direction from the other side more than we seemed to consciously recognize. An idea can be there for us, and we can perhaps even purchase equipment knowing that something is going to take place, but the equipment will remain idle until some "final" ingredient comes into place. Since we are hard workers and we are also curious about things when we do not follow through with an idea, it is usually because of an inner knowing that the time is not yet right.

After we had moved into the house in Reno, and were becoming accustomed to the work of the AA-EVP, we attended a gathering of a few EVP and ITC researchers that was sponsored by the Noetics Institute.[64] At the gathering we met Professor Euvaldo Cabral Jr,[59] who suggested that we place the camera within four or five inches of the television screen and focus it past the screen so that the resulting feedback produced a chaotic texture on the screen. That was the last bit of information we needed to begin experimenting with Video ITC.

In November of 2001, we were about to conduct an EVP experiment when we finally decided to try to do a Video ITC experiment. The process that we used is explained in Chapter 13 if you are interested in duplicating our experiments.

In preparation for the experiment, we dug out our video camera and a camera tripod and placed the camera within inches of a television set. Then, we darkened the room for the experiment by covering the window with a blanket, but we kept a sixty watt desk lamp on so that we would not knock over the camera. After the experiment, we transferred the audio from the EVP experiment into one of the computers in our office area. The first part of the experiment we reviewed was the audio, in which we heard a voice saying, *"Continue with the pictures, I will help."* This was very exciting and so we immediately loaded the video into the computer and began viewing the video, frame by frame. The first lesson we learned is that it is a very slow process and takes more time to review than does an EVP session. There was nothing on the video that we could find but we could see that we may have had the camera focused incorrectly.

The next day we conducted another EVP session. First, we thanked those on the other side for their assistance and then asked who had said that they would help with the pictures. The answer came immediately after the question, *"Anthony."* The name was not one we knew, but we were nevertheless encouraged that someone on the other side might be taking special interest in our work.

In late November of 2001, we ran another video experiment. As usual, we played the piece of music that we always play while setting up the equipment. Everything was as it had been on the first experiment. The room was darkened, we used the same equipment and we meditated as before. The only parameters we changed were the zoom and focus of the camera. That, we hoped, was the problem with our first experiment.

When reviewing the ITC session, we did notice a couple of features that looked like faces when we saw the video frames in the review window, but they were very dark. Nevertheless, that moment was breathtaking. There was actually something phenomenal in our video! But, our excitement quickly turned to frustration as we "grabbed" one likely frame after another and tried to make the features clear enough to tell if they were really faces. Most remained little more than a hint of a face, as if we were seeing a face through a very foggy window. This was also our first introduction to the frustrations of trying to print a feature on paper. Those that were a little sharper and were clearly a face, as seen in the photo editing program, printed as blobs of gray and black with our laser printer or equally useless blobs of color in our ink jet printer. There was nothing we could do with our first, triumphant examples of Video ITC outside of our computer.

In early December, we sent two of our most promising video frames from our second experiment to Erland and Mary Babcock as an attachment to an email. Erland is probably the most knowledgeable person about ITC pictures in the United States, and we were fortunate that he was a member of the AA-EVP. He wrote back that the pictures were too dark to print but that he would try to work with them. He said that both he and Mary saw a man in one of the frames. Interestingly, what they saw was not what we had been focused on in that frame. The feature that we were focused on was the bust of a man that nearly filled the frame.

What Erland and Mary did with the frame, and what they sent back to us, was thrilling because just to the left of the major feature we had focused on was the face of a man wearing a black hat. The picture was difficult to see, but for us it was thrilling because we could see the hat, and the man looked like he had a beard. It might not have been the greatest but we had gotten our first ITC picture with the help of Erland Babcock's experience in working with video. With Erland's guidance, we could see that there might be a possibility that we could do further work in Video ITC and obtain results. It had been our dream for years to work with Video ITC and that dream has finally been realized. It took a long time to get to sleep that night.

More excitement was to come a few days later. Erland had tried to print the frame that had the man with the wide brimmed hat. When he took the picture off of the printer, he found something quite different. He emailed us that he did not know where this picture came from. He even discussed the possibility that it had paranormally appeared when printed. He wrote, "Wait until you see this image; it looks like some-one's grandfather, complete with a handle bar mustache."

The picture that Erland sent us through the mail was a true gift from Spirit. That first picture, which we affectionately know as the "Standing Man," hangs on a bulletin board in our experiment room. His hair and his mustache can be seen, he is wearing a suit jacket and what looks like an ascot is visible around his neck. One of his hands is also visible and he may be holding a cane. The Standing Man is in-cluded in this book as Figure 12-1 in Chapter 12, but if you are able, we invite you to look at the color version of this image at http://aaevp.com/resources/pictures_in_no_dead.htm.

Where did the Standing Man come from? Was it paranormally placed on Erland's printer? After seeing the image, we once again brought the original up on our computer screen. We found that the Standing Man was oriented at a minus ninety degrees on the frame and just under the beard of the man with a hat. The pinkish field that is the legs of the Standing Man can be seen in the color version of the man with a hat. Our spirit team had shown us with the first frame we con-centrated on, that the features were not going to occur in just one ori-entation. With EVP there are often messages on the reverse side of the tape. With video ITC you have four directions per frame, and at least with us, the images will appear on any of those four orientations.

As it turned out, the discovery of the Standing Man also helped us with critics who appeared with our introduction of these features in the AA-EVP NewsJournal. It was immediately claimed that we were only picking up stray television signals, even though the television had no antenna, and even though using the Video-In connection of the televi-sion effectively disconnects it from the receiver. Our answer to the critics was to ask when they had last seen television programs with standing people oriented at ninety degrees from vertical or upside down. Our transpartners had helped us with answers in anticipation of these criticisms.

Reviewing the video frames took much longer in the early days of our Video ITC experimentation. It was usual for us to take consider-able time with each frame as we enlarged them, rotated them in ninety degree increments, and adjusted the contrast and intensity to bring out every last detail. There was also a period of time in which we did not know what to think of the phenomenal features we were finding.

Every possible alternative explanation was considered. Tom is an electronics engineer, and after much thought on the subject, he decided that there was no explanation for the presence of the features that was based on known physical principles. After all of this, we were forced to accept that the features were, indeed, paranormal in origin and that they truly represented a form of spirit communication. Video ITC features were pictures of "dead" people. How could that be? Why were there faces in the video noise and how were they formed?

Phenomena in our Lives

In early 2002, our experiments were put on hold as Tom had angioplasty. It was a stressful time for us but we were glad that the problem could be corrected before serious damage to his heart occurred. The day after he was released from the hospital we celebrated by conducting an ITC experiment.

During experiments, we operate several tube-type AM radios as sound sources. The same piece of music is played each time as we set up for the experiment and meditate. Then, of course, there is the television set and camera power adapter. So when we are finished with an experiment, we manually turn everything off. Power is also removed from the equipment power cords by turning off power to the strips we use for surge protection and power distribution. Finally, the power switch at the door is turned off, removing power from the wall socket that supplies the power strips. These measures are taken because the power switches are there and not for any safety or paranormal reason. The point is that all of that equipment is really turned off when we are finished with an experiment.

After the experiment, we took the sound and video tapes to the office area to be loaded into the computer. It is cold in Reno, Nevada, in February and we routinely go through the house late afternoons to close the window blinds to retain heat. The experiment room had been entered two times after the experiment and there was nothing out of the ordinary. All of the equipment was turned off.

That night, we ate dinner at around seven. Lisa recalls what took place, "Carrying my dishes to the kitchen, I heard noise coming from down the hallway. There is only the two of us living in the house, so that was rather shocking. I called for Tom and we headed off down the hallway. Believe me I was in the rear with Tom taking the lead. As we approached the experiment room, we quickly realized the noise was coming from the compact disc player. The player is used to play our meditation music, and occasionally, the Portuguese crowd babble disc that Sonia Rinaldi had provided for background sound in EVP experiments. The babble disc was playing!"

The player holds three compact discs. The first selection is the meditation music and the second is the babble disc. Not only did the light switch that controls the equipment have to be turned on, the associated power strip for the player and the player itself also had to be turned on. Further, the compact disc player had to be selected because the unit always turns on with the radio selected first, and the second disc slot had to be selected because the unit always begins with the meditation music. The whole event was just impossible! By the way, we are the only humans living (in the flesh) in the house and the cat has never turned on lights or power strips and has never showed an ability to play compact discs.

In shock, and trying to logically figure out how this could have happened, we went over every bit of equipment and every power switch. Finally, we concluded that there just was no logical explanation. Our invisible friends had just told us that everything was going to be fine with Tom and that they were very glad he was back and running experiments. The phenomenon was simply a big welcome home. Our next thought was how shocking it would be if the babble sound was to suddenly turn on in the middle of the night while we were sleeping. There are limits to our search for phenomena.

Another strange thing happened a few months later, but it could not be as easily tied to the experiments. Two apports were received at a presentation that we made in Toronto, Canada for the International Spiritualist Federation. These were small sticks restaurants use to tell the customer how their steak has been cooked. These sticks said "Medium" on them and one of them was actually apported into Lisa's shoe as she was walking. The "medium sticks" are something of a trademark of the Imperator Group who communicates from the other side via independent letter writing and apports with the Society for Research in Rapport and Telekinesis.[25]

What we Find in the Frames

Our first successes with Video ITC brought the concern that it all might just be a fluke and that the next experiment might not produce phenomenal features. Thankfully, this was not the case. It is true that some experiments produced better examples of Video ITC than did others. This was partially our fault, as the focus and zoom settings seem to be critical and we are not able to lock these in a particular position. These settings are reset each time we turn off the camera. Various camera presets were tried such as "Sun and Sand" and "Sports," but we always returned to the portrait setting. The small images that we were getting were frustrating, and after six months of experiments with the camera just inches from the screen, we were directed by our spirit team to move the camera back. The results of doing this immediately pleased us, as we found that we were still recording faces but many of them were larger and it was not necessary to increase their size for viewing. Moving the camera back allowed for a different kind of detail without the interference of the many pixels.

In corresponding with other researchers, and based on their comments, we experimented with placing the camera at different angles in relationship to the surface of the screen. As of the writing of this book, we continue to have the greatest success with the camera three feet from the screen and in the "portrait" preset. The camera is set up in three different positions and we record in each position for ten seconds. These positions are straight in front of the television and then at about twenty and forty degree angles to the right of the television. In an effort to improve the features, we were guided to use a sixty watt lamp pointed downward, right under the television screen. Through later correspondence, we have learned that other researchers have had good success with a sixty-watt lamp, but this was unknown to us at the time. One researcher aims the light at the video camera and feels that this has improved his images; another researcher aims the light at the television screen with good results. Experiments in which we aimed the light at either the camera or television were disappointing. This shows how many different ways there are to conduct Video ITC experiments and it also shows that one method might work for one experimenter and not for another. Some ideas on how to conduct your own video ITC experiments are explained in Chapter 13.

From the beginning, we had received pictures of people from various periods in time. A trip to the library for costume books led us to believe that most of the clothing and headgear we were seeing were worn from the sixteenth through nineteenth centuries. Other images collected were of people that we would consider being extraterrestrials. Some of them have appeared as they are portrayed in the media, but some are clearly ones that we have never seen. One entity that we have seen more than once in our own images has a point on its head and additional points where our ears would be. The three points are about the size of a large pointed ear. Erland Babcock had previously collected a similar image, and we found another that has been received by Alfonso Galeano of Barcelona, Spain. Alfonso's example can be seen at http://webs.demasiado.com/ufonews/ as "Psicoimagen de un supuesto DUENDE."

Many animals, which are often in close association with people, have been received in our experiments. For instance, we show in Figure 9-1, a young man in uniform proudly holding his little terrier dog in his arms. Many scenes appear to be of groups of people indoors and there are many that show people in the countryside. The frames that appear to have scenes are not sufficiently detailed to print, but we will attempt to post a few on the AA-EVP website.

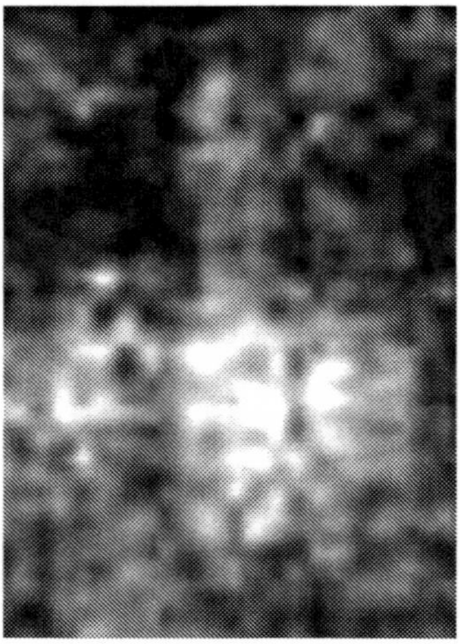

Figure 9-1: Man with dog

After a few phone calls, we developed a wonderful long distance relationship with Larry Dean and Patricia Begley, who live in Great Britain. As a spiritual healer and trance medium, Larry was interested in conducting Video ITC experiments and called when Tom was in the hospital. Lisa recalls, "I was so upset at the time that Larry called. He is such a kind and caring soul and his timing could not have been better. He was easy to talk and there were so many common interests. He helped me stay upbeat and look for a positive outcome when Tom was in the ICU." Since that first call, the friendship between all of us has grown.

Larry and Patricia were sent a draft of Chapter 12 and 13 from this book to see if the Video ITC instructions would help them in getting ITC images. Soon, Larry was calling; excited because they were recording Video ITC features. Larry and Patricia have been setting the camera back only three inches from the television screen and unlike the small images that we obtained up close, most of their images cover the full frame.

Figure 9-2: Tom's father (left) and ITC feature resembling his father

One thing that Larry and Patricia do before an experiment is to ask for a specific person to come through. Using this approach, the two researchers have collected features of several friends and relatives, some of which can be seen at aaevp.com. In December of 2002, we called on Tom's father and asked him to appear in the middle of a frame so that it would be easy to find him. He did just that, as you can see in Figure 9-2.

Early in our experiments, we collected a feature that looked a lot like Albert Einstein, the famous scientist who has been occasionally associated with ITC research. The picture was not clear enough to be definite, but we were given the information that it was Einstein. So we hoped that he might be working with us.

Information was being received during the EVP experiments and on the audio portions of the video that seemed to be an acknowledgement from the entities that they knew we were conducting experiments and that they were working with us. Late in the year, we began receiving numbers on the audio that could be perceived as dates. For instance, in December, the audio on the camera had a voice that said, *"twelve-twenty."* Could this be December 20? An experiment conducted on that date provided features that did seem to be clearer. As of the writing of this book, numbers on the audio have only been received three times, and all have led to be rewarding experiments.

An experiment on March 1, 2003, turned out to be particularly exciting. While we were going through the video frames that had been downloaded into the computer, Lisa exclaimed, "That looks like Friedrich Jürgenson! Look, he is holding some kind of little animal next to his face." As she relates the rest of the story, "I ran to another room to get his picture, and when I approached the computer screen, all of a sudden it hit me that the little animal was a bird!" Unfortunately the image is not one of our better ones. As you may recall from the chapter on the history of EVP, Friedrich Jürgenson first recorded EVP while recording bird songs, and in his lifetime, he was a dedicated worker for the protection of birds. When Tom first saw the Jürgenson picture, he looked at the bird and declared it to be that of an owl. At the time, we did not put any significance on it being an owl. Later, we remembered that the first thing Jürgenson got on tape was something about nocturnal bird songs. An owl is of course nocturnal. This all seemed to add confirmation to the image. Little did we know there was more to come.

Cross-Correspondence via Transfiguration

Later that March, we had an appointment to see Jean Skinner, a British transfiguration medium who had been invited to the United States by the Golden Gate Spiritualist Church in San Francisco. As the Directors of the Department of Phenomenal Evidence for the NSAC,[9] we felt that it would be a good idea to drive from Reno to the Bay Area to attend one of Jean's sittings. There were to be a few small sittings in which there would only be five people sitting with the medium. A large demonstration by the medium was also planned. There would be fifty people attending that sitting. The five-person sittings were already booked, and the only available opening was for the larger session. Talking with Sonny Gee, the President of the church and the person who had been responsible for bringing Jean Skinner to the States and who was coordinating the sessions, we requested that he contact us if there were any cancellations for a five-person session. It seemed fate had helped us when Sonny called and told us that a couple attending one of the smaller sessions had cancelled. This would allow us to see the medium up close.

The séance took place in a small room, with the windows covered and with three red lights for illumination. The illumination was very good and it was easy to see the medium and what was taking place. After instructions to the sitters from Jean's assistant, Jean allowed herself to settle into a deep trance. After maybe five minutes, Jean's Asian guide spoke to us. The guide called on Tom to sit in a chair that had been placed directly in front of the medium. Tom did so, and after just a few words, Jean's guide requested that Lisa also come forward. Together, we held Jean's hands as she spoke a few words and then allowed herself, through her guide, to become receptive to other personalities. Several of our relatives came through in quick succession. Jean's spirit team is working to develop an ectoplasmic voice box through which emerging personalities might speak to the sitters. Tom saw an electric blue glow near Jean's left shoulder and neck, and occasional blue wisps of energy flow from her mouth, but there were no spoken messages while the personalities came through. Following the assistant's instruction, we spoke our confirmation or told Jean that we did not recognize the personality after each one came through. Instructions were given to us to make comments, such as, "The person I think is coming through had a mustache." In which case, if we did have the right person, evidence of an ectoplasmic mustache might begin to form. When we recognized someone, we spoke that person's name. If that person was truly present then the entity's body language, such as a nod of the head, would show us that we were correct. If we were wrong, the entity would pull back.

Sometimes, the personality silently greeted the sitter with heart-felt expressions, as if a long lost grandmother was pleased to once again be before a favorite granddaughter. All of this was witnessed as changes in Jean's facial expressions; these were accompanied by subtle changes in what appeared to be an ectoplasmic mask.

While we sat in front of Jean and her guide, we watched as a new personality transfigured Jean's features. Lisa was the first to recognize who it was. She said, "Can this be, Friedrich? Friedrich Jürgenson, is that you?" The face nodded and leaned toward us. After thanking him, we told him how honored we were to have him appear. Lisa said, "You sent us a picture." The personality of Jürgenson quickly nodded his head.

For whatever reason, Lisa next said, "What we want to know is what happened to Timestream?" Jürgenson sat up straight, nodded, seemed excited and then the face melted away and another appeared. Lisa was almost speechless but managed to say, "Konstantin Raudive? Is that really you?" The personality with a decidedly rounder face acknowledged that it was. Tom said, "You forgot your glasses," and we could see dark rimmed glasses trying to form in the faintly seen ectoplasmic mask. Others in the group exclaimed, "Look, you can see glasses forming!" Then Raudive was gone and yet another figure began to form. It was Tom who first recognized this new personality, saying, "Dare I say ... can it be (disbelief and shock) ... Einstein?" The personality acknowledged and we were speechless! Our minds went blank from what had just taken place with these three men coming through. Einstein looked at the medium's companion and she almost shouted, "It is him! It is Einstein!" He looked around the room for a moment as he nodded to the other sitters. None of us had the presence of mind to ask a question. He looked at each person once again and then was gone.

Experiencing nonphysical phenomena and the energy that accompanies such extraordinary events is one of our favorite things. However, the pursuit of phenomena has other importance for us that far exceeds the joy of the experience. Nonphysical phenomena offer important evidence that we survive physical death. These phenomena also offer tantalizing hints about the operation of reality. No field of interest in the paranormal is isolated from the other fields. Transfiguration phenomena, such as that so ably demonstrated by Jean Skinner, has a direct correlation in EVP, as does EVP echo the principles which underlie telekinetic phenomena. Learning to understand something of one aspect of these phenomena often leads to further understanding of the others.

As we must close this chapter on our own ITC experiments, we know that there will be more exciting things that we wished we could have included. In the past few weeks, our transpartners have sent us the clearest image to date. It is of a lovely lady. Bless Sarah Estep, we were so pleased with the image that we emailed her a copy. She has always been the most wonderful support to us. She wrote, "She is *beautiful* and one of the very clearest I've ever seen anywhere. She's different from most spirit pictures, in that she looks like she still has 'living' within her—has not passed on and looks like a spirit. She also has a lovely, colored dress on her, and I think a hat." Again, we apologize and say that these images just do not print clearly, but we have tried to put the black and white version of her as Figure 9-3. The original color image can be seen at aaevp.com.

Another recent occurrence is a person speaking at the beginning, and often, at the end of experiments. He speaks on the audio of the video and calls himself simply, "The Assistant." Other voices on the audio are not as easily understood. Messages such as, *"Fading on amplitude"* and, *"Getting light scope,"* are a sample of the kind of things about which we cannot even guess as to their meaning.

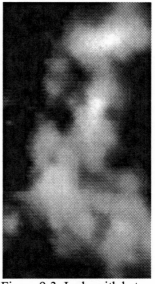

You may wonder why we are so interested in establishing EVP and ITC contact with particular helpers on the other side. Samples of these phenomena can be collected with little more than the willingness of an entity to speak or appear in our video and still cameras. However, we also know that some of the most important work that has historically been accomplished in this field, has been accomplished with the

Figure 9-3: Lady with hat and dress

assistance of a group of entities working together on the other side in an effort to establish a major communication bridge across the veil. For now, even fuzzy ITC images are considered an enormous gift and a small view into other dimensions of existence. Each EVP voice is cherished and we are ever grateful to those on the other side who take their time to work with experimenters all over the world. Without them, there would be no EVP voices and no ITC images.

Part II

Experimenting with and Understanding EVP and ITC

EVP and ITC are more technologically oriented than are other forms of spirit communication, and for many, working with electronic equipment to communicate with discarnate loved ones is more of a challenge than sitting down to speak with a medium. However, remember that the technology you will need to use is new only once. You really do not need to become any more technically involved than what is required to record EVP and this is covered in the first part of Chapter 10. If the simple step-by-step instructions offered here are followed, and only one process is mastered at a time, you will soon be an "old hand" at collecting these phenomenal messages for yourself.

As for the theory concerning these phenomena that is discussed in some of the following chapters, the theories are included for completeness, and for the benefit of those who are interested in developing devices and new hypotheses. It is by no means necessary to understand them in order to successfully collect EVP and ITC messages.

Chapter 10

Recording EVP

In describing the history of EVP and ITC, and the experiences people have had with these phenomena, we hope that we have whetted your appetite enough that you are now interested in learning how to collect examples of these phenomena for yourself. In this chapter, there are step-by-step procedures for recording, followed by more in-depth information on recording.

What is offered here are the techniques for EVP experimentation that we have found to work for us. The different ways people have adapted the "standard recording technique" to suit their personal needs and available equipment is impressive. So, while we recommend that you begin with the basic approach we describe here, we also recommend that you experiment to find variations of this technique that best suit your situation.

Recording EVP in Controlled Conditions

By "controlled conditions," we mean that you will be recording in your home, or in other circumstances in which you have the ability to control background sound and recording conditions. There will be certain pieces of equipment that you will need. You probably already have some of the equipment around your home, so try to use what you have to begin with.

When we first began experimenting, we found that we had all of the necessary items. Equipment purchases were not made until we knew that we could obtain the voices and were sure that we wanted to continue this pursuit. Sarah Estep mentions beginning with equipment that she had that only half worked, but she still recorded voices from the invisibles.

Basic Equipment

Recorder: EVP has been recorded on all types of equipment. If you use a cassette recorder, it is best to find one with mechanical controls that allow easy, repeated review of the voices. Be sure the recorder has a counter. The personal note recorders, known as IC recorders, have been on the market several years now and we believe that there is sufficient evidence to indicate that these recorders perform better for EVP collection than do the cassette recorders. The IC recorders are about on a par with reel-to-reel recorders.

Microphone: A microphone will help when making a record of your comments during the recording session, and allow you to introduce external sound sources. EVP messages will have more meaning if you have a record of the questions asked and the recording circumstances.

The use of an external microphone is recommended with cassette recorders because the built-in microphones tend to pick up motor noise. The IC recorders do not generate mechanical noise, so the external microphone is not considered necessary for these, unless you desire to have better sensitivity.

Headphones: EVP voices frequently are not loud and many may be missed unless headphones are used for listening to the recordings. The earmuff-type, which completely covers the ear, is usually used.

Tape: Any low noise, high sensitivity tape may be used. Sixty-minute tape (thirty minutes each side) is recommended. Audio tape does produce a certain amount of "hiss" noise that is not useful for the formation of EVP and can obscure the voices. However, sound editing software can usually remove this sound. There are also hiss reduction hardware components that can be purchased for this purpose if you do not have a computer.

Speaker: A separate speaker is not necessary but is good to have. With the speaker, if there is a Class A EVP, you will be able to play it over the speakers so that everyone in the room will be able to hear the message.

Noise Source: The following sound sources have been found useful as background noise for EVP experiments:

♦ An AM radio that you can tune off-station.

♦ A fan

♦ A short-wave radio

♦ A cassette or CD player so that you can play specifically pre-pared sound tracks, such as crowd babble or foreign language. A radio to play a foreign language broadcast is also an option.

♦ A source of running water such as a fountain.

Preparation

Meditation and Prayer: Always tape when your energy level is highest! Before conducting an experiment, take time to become quiet. Still your mind and focus on reaching those now in other worlds. Conduct a short meditation and/or prayer and ask for those who exist on the other side to help you create a bridge to the other side.

Scheduling: Entities will speak on tape at any time of day or night. In the beginning, however, it is advisable to record at a regular time and place. By doing this, the entities learn when you will be recording. After making a connection with the entities, you will be able to collect EVP at any time and in any location. Try to find a place that will be quiet and free of interruptions. Background sounds are all right, but it is important that you are aware of these so that you will recognize what sounds are natural and what sounds are EVP. Keep your recordings short. You will want to listen to each part of the recording very carefully and this can take time.

Background Sound Source: The entities use sounds in the environment to help form EVP messages. Most recording situations have some background sounds, but you may wish to add sound to your recording environment. Begin by using any of the sound sources we have previously mentioned. Also, the communicating entity will sometimes remodulate your voice or other sounds in the environment. This is one of the reasons experimenters will sometimes use a foreign language sound source.

A good rule of thumb: The quieter the electrical circuit in the tape recorder, the more background sound you will need to supply.

You will find that experimenters try all sorts of devices and energy sources to help the entities communicate. You can find ideas in past NewsJournals in the AA-EVP Archive[1] or via a search on the Internet, but in the end, let your intuition be your guide.

Recording: Vocalize your comments during an EVP session. Many experimenters begin with a short prayer and an invitation to friends on the other side to participate in the experiment. It is helpful to begin an experiment by speaking your name and the date.

The entities will often come through as soon as the recorder is turned on. These beginning messages are often the loudest, so it is a good idea to turn on the recorder and wait a few seconds before announcing yourself and then ask the first question. Your questions should be recorded, and you should leave a period of time between each comment for the entities to respond.

Some experimenters make an "appointment" with the intended entity the day before, during prayer or meditation. Some also provide verbal feedback, about how successful the previous session may have been, before the session so that the entities will know how the last experiment went. The feedback need not be recorded, just go to the experiment room and talk to the invisibles as you would a good friend. Some experimenters place paper with written questions in the EVP experiment area the day before. Several experimenters do this and say that the entities can read these and respond accordingly.

By the way, it is not necessary to record in the dark. In fact a humorous EVP that we collected when we experimented with recording in the dark was, *"This is not a séance!"*

Playback: The paranormal voice is not usually heard until playback of the tape. The voices may speak in whispers at first, but experimenters report that the voices tend to become stronger and clearer as the entities gain in experience. Voices may not be recorded in every session and it may take several sessions for you to discover the first voice. Hearing the voices is something of a learned ability.

Keeping a Log: It is helpful to maintain a written record of recording results. Include the date, time, place on the counter or file system where the message is received, the message itself, and the question asked. Speaking your name and date at the beginning of a tape recorded session is also helpful; however, this information can be stored as a file name for recordings that are stored as computer files. Be sure to label recording tapes. Experimenters report that weather may affect results and the aaevp.com site has geomagnetic and solar reports for this purpose. There is also a link for moon phase information.

The Recording Session: Begin by turning on your equipment and conducting a test by recording and playing back a few seconds of sound. This will assure that you have microphone and tape recorder properly arranged and your sound source at a level that will not drown out your voice.

1. Take some time now to relax and focus your attention on what you are about to do. This is a good time to seek your "center" or that meditative attitude that best helps you clear your mind of external influences. It is best to turn off the sound source during this time. Perhaps you can play some meditation music instead.

2. After you have finished the meditation, turn the sound source back on. If you were playing music, turn it off. Speaking out loud, announce to the entities that you are about to turn on the tape recorder. This is a good time to explain why you are there and what you would like from the entities.

3. Turn on the "Record" function on the recorder and wait ten to fifteen seconds.

4. If you intend to maintain an oral record of your experiments, identify yourself and say the date.

5. Ask the first question and wait fifteen seconds or more before asking the next question. It is a good idea for the last question to be an invitation to the entities to say something of their choice. Keep your recording short, no more than five minutes is best.

6. Tell the entities that you are about to turn off the record function and thank them for helping you in your experiment. Wait ten to fifteen seconds before turning off the recorder so that they may give you any last minute messages.

7. Shut down your equipment and begin the process of reviewing the sound track for that session.

These steps are all there is to an experimental EVP recording session. If you plan to use an IC recorder for EVP, seriously consider reviewing the sound track in a computer. If you do not have a computer, then use a good set of headphones.

Remember also, that you can use a computer as your recording device.

Recording EVP in Field Conditions: Recording in "field conditions" is defined as recording in a situation under which you have little or no control over environmental conditions. For example, the opportunity to record for EVP in a purported haunted location.

Special considerations: A list of things to consider for field recording is provided in Chapter 7. The main differences between field and controlled experimentation are identified in that list.

Things You Should Know about Recording EVP

It is important to understand that working with EVP and ITC is actually very simple. While we have described the usual technique for doing so, remember that there are probably as many different ways of experimenting with EVP and ITC as there are people experimenting. Actually, anything that records can be used to gather EVP, so there are numerous possibilities in what experimenters might decide to use to record these paranormal voices.

As we have said so many times before, the experimenter is clearly part of the recording circuit, in that the experimenter has something to do with the quality, quantity and nature of the collected EVP simply by being who they are. That is why some people will tend to be more successful than are others, even though everyone may be using essentially the same equipment and technique. A few theories as to why this is true will be discussed in Chapter 11, but for now, remember that your attitude seems to be very important. If you are enthusiastic, and have a strong sense of desire to succeed, then you are more apt to be successful with EVP.

A few well respected researchers are so convinced that EVP collection is dependent on the experimenter that they reject what Paolo Presi[31] refers to as the "Radiophonic Model." Like Presi, these researchers believe that what is used to record and how it is used, is not particularly important because it is the person involved that is the determining factor. This view is supported by evidence that especially designed devices that have worked very well for the inventor, have not proven better for EVP than the usual audio recorder when used by other experimenters.

Recording Entity Voices

First, remember that the recording equipment need not be expensive or elaborate to work for EVP collection. Rather than the quality of the equipment, your ability to hear the EVP should be the first concern. Assuming that you have normal hearing, or that you can hear the EVP when you amplify the sound track, then the main challenge will be learning how to distinguish EVP from the many other noises often found in sound tracks.

Raudive and Jürgenson both said that hearing EVP was something that had to be learned. With this in mind, it may be a good idea to spend some time browsing the Internet and listening to the many examples of EVP found by conducting a search for "EVP." Go to the AA-EVP website at aaevp.com and listen to some of the examples there. Also, the AA-EVP has a members only discussion board that is very helpful for those just beginning to experiment with EVP. Members share their EVP samples with each other, thereby receiving immediate feedback about what others hear. Ideas on equipment, different ways of recording and using sound editing equipment to hear EVP are also discussed on the board.

EVP messages have a characteristic cadence and are often short but meaningful. It is sometimes possible to recognize who is speaking by the sound of the voice. It is even possible to tell if the speaker is male or female, an adult or a child. Sometimes, the voice will have a mechanical sound, as if a computer or electronic voice box is generating the words. The messages will tend to have a logical beginning and end. By comparison, if voices from a radio station were recorded during an experiment, the words would often begin or end in the middle of a sentence and would normally be unrelated to questions asked.

EVP messages sometimes trail off at the end, as if the speaker is rapidly running out of energy. Also, mundane sounds can be transformed
into an EVP message. For instance, the barking of your neighbor's dog
might become something entirely different on the recording. Your own
voice will sometimes be remodulated or changed into words that you
did not speak.

Hearing and recognizing EVP may be something of an art. Yes,
some EVP, the ones described as Class A EVP, are sufficiently clear
so that there is little doubt as to what is said. Such EVP are decisively
evidential. However, most beginners do not immediately collect Class
A voices. In Class C or B EVP, you may think the EVP sample says
one thing, but other people may hear a different message. This is especially true if the witnesses are not accustomed to hearing EVP. It is
important that this does not discourage you. As you continue to experiment, you will build a bridge to the other side with the cooperation
of the entities who agree to communicate. With this assistance, the recordings will improve in both the quantity and quality of the phenomenal messages.

Recording EVP takes enthusiasm, patience, perseverance, good
hearing and an open mind. Anyone with these qualities can learn to
collect EVP.

More Information on Equipment

There are no rigid rules for experimenting with EVP. Anyone can expect to succeed in recording EVP messages if the instructions at the
beginning of this chapter are followed. However, once you have succeeded, then it might serve you well to try variations to these instructions, using your previous success to help you determine if the variations help.

Audio Recorder: EVP has been recorded on just about everything
that will record voice frequencies. As was noted in the history of EVP
and ITC, the first EVP recordings we are aware of were made on a device that scribes audio soundtracks onto wax. Sarah Estep still uses a
reel-to-reel tape recorder with great success and we began recording
with a reel-to-reel recorder as well. When our reel recorder quit working, we found it difficult to replace and so progressed to several different cassette recorders. However, we were very impressed with our first

use of an IC recorder and now prefer it to a cassette recorder. Experiments we have made with two different IC recorders have produced very good messages. People also commonly find phenomenal voices on their telephone answering machines. Many people successfully collect EVP directly into their computer—sometimes without a microphone.

Cassette tape recorders should have mechanical controls that allow easy, repeated review of the voices. The recorders that have "piano key" style controls are easiest to use. It is a common practice, when reviewing the resulting sound track, to "rock" the tape back and forth many times to listen to short segments of the tape. Be sure the recorder has a counter, because you will want to be able to return to a specific portion of a tape. A counter will also help you maintain good written records about where the EVP is located on the tape.

Veteran experimenters recommend the use of component-type cassette recorders because of their higher quality circuitry. These units require an amplifier and an external microphone. However, more background sound may be necessary if the internal circuitry of the unit is very quiet. By comparison, most portable cassette recorders have more noise in their circuitry and they tend to have more tape hiss, which can obscure the phenomenal voices. The biggest drawback of most portable tape recorders is the need to use an external microphone to avoid picking up noise from the cassette drive motor.

IC Recorders: The tape recorder of choice for early EVP experimenters was the reel-to-reel tape recorders. These had excellent, quiet sound tracks and it was easy to play the reverse of the sound track by twisting the tape. As a rule, reel recorders also used vacuum tubes rather than transistors. As the cassette tape recorder dominated the market, it became more difficult to find parts for and maintain the reel recorders. The cassette recorders were generally based on transistor technology, were very portable and relatively inexpensive, effectively obsolescing reel recorders for all but the most particular audiophiles.

The digital note taker, or IC recorder, has arrived on the scene in recent years. The early models lack the recording quality found in the cassettes, and certainly in the reel recorders, but they are convenient! They are not particularly expensive, they will fit in a shirt pocket and they weigh only a few ounces. The electronics in the IC recorder is transistor based, but we believe they are of the low power Field Effect Transistor (FET) variety, as opposed to the type of transistor used in larger battery, or AC powered equipment.

In a test conducted by the Delaware Valley Demonology Research,[78] the reel and IC recorders were rated highest with a "10" for EVP collection while the standard cassette recorder was rated a "7." This is consistent with our observations because we have noticed that reports of EVP have seemed to decrease as the cassette replaced the reel recorders. However, we have also noticed that the reports of EVP have greatly increased as certain models of the IC recorders came onto the market. This must be speculation because we do not have all of the necessary information about the technology used in IC recorders. Nevertheless, we believe that the FET performance characteristics are very close to that of the vacuum tube and rather different from the standard transistor. Even today, some of the most outstanding EVP and ITC are being collected by people using vacuum tube devices. The message seems to be that EVP and ITC experimenters should look for FET or very low power devices for experimentation.

The Panasonic RR-DR60 IC Recorder is legend amongst AA-EVP members for EVP. As the story is told, soon after the product came out—we believe it was the first of its kind for Panasonic—customers began returning them because they were finding voices in the sound tracks that they had not recorded.

Other brands and models of the IC recorders have been used for field recording of EVP with great success. With that said, the later models of the Panasonic, the DR60, QR-80, QR-100 and QR-200 have all received praise from various AA-EVP members. These Panasonic models are no longer manufactured, so it is necessary to look for them in the used market.

It is not a reasonable assumption to expect that the newer models of Panasonic IC recorders will be equally effective for EVP. At the time of this writing, there is too little experimental evidence to indicate one way or the other. If the FET technology is easier for the entities to work with in the formation of the messages, then newer models should work as well as the early models. We know that, as the newer models are able to record higher quality sound tracks, which will mean that they have a quieter electronic circuit and higher sample rates, they will probably require more background sound for the formation of the phenomenal messages.

Tests on various IC recorders also conducted by the Delaware group[78] show that some types are less effective for EVP than are others. The cause for this difference seems to be in the sample rate. IC recorders with a sample rate that is less than twelve kilohertz are said to work the best. However, finding the sample rate of a particular recorder may not be easy. The DR60 and QR-200 that we use were previously known to work well in the collection of EVP. From Panasonic, we have learned that the RR-DR60 has a fixed sample rate of six point four kilohertz. The RR-QR80 and others in that series have a sample rate of eight kilohertz in both Slow Play (SP) and Long Play (LP) modes, and sixteen kilohertz for the High Quality (HQ) mode. For EVP, we recommend the SP or LP modes with a microphone sensitivity of high and with the voice activated switch enabled.

The Panasonic DR and QR models are actually very poor recorders for other purposes when compared to devices that are designed for music. This is probably because the voice-activated switch tends to turn off and on for no obvious reason in some models, and the internal compression and threshold settings seem to be constantly "hunting" for a null point. However, it may be the combination of the FET circuitry and the abundant internal, electrical noise that makes these devices so effective for EVP collection. The recorders also have a little speaker that cannot properly reproduce the voices. Thus, it is best to always use a headset for reviewing sound tracks for the paranormal voices, or download the recordings into a computer so that you can use an audio editor such as Adobe's Audition.[18]

IC recorders have a recording session or file indexing system that provides a record of session number, time and date of the recording, and duration in minutes and seconds. This information makes it possible to quickly access one of many sound tracks. If the sound tracks are transferred to a computer as individual audio files, it will be necessary to label the sound file to agree with the log entry you created for that experiment. This is also true of sessions recorded with a cassette recorder.

The future history for IC recorders and the FET technology has yet to fully unfold, but it is already apparent that this new technology will continue to improve EVP experimentation. The aaevp.com website will be updated on this subject from time to time, so check the site to see what develops.

Should you decide to purchase an IC recorder for EVP, remember that you will need a computer and sound editing software like Audition[18] to support the recorder. Some information on how to download recordings into a computer, and the use of sound editing software, is given in Appendix B and there is also information on the aaevp.com website.

In the case of devices such as telephone answering machines, we also recommend that the sound track is transferred into a computer for review with a sound editing program. A handy gadget for transferring sound tracks from an answering machine is an inductive coupler. Radio Shack sells them as a "Telephone Pickup," Part Number 44 533.

Microphone: If you are using a cassette recorder, a microphone will help you make a record of your comments during the recording session and enable you to introduce external sound sources. Experimenters have found that their EVP messages often have more meaning if they are accompanied with a record of the questions that were asked and the recording circumstances.

In a test conducted in behalf of the German ITC group, the VTF,[53] microphones were connected to a stereo recorder as channel A and B. One microphone was covered so that it could not pickup sound while the other was left open to the environment. In tests using this configuration, EVP were collected on the channel connected to the uncovered microphone and not on the channel using the covered microphone. At first glance, this would seem to suggest that the EVP is injected into the circuit acoustically, before it is converted to an electric signal. However, we believe that this test may also confirm that background sound is required for EVP, since the covered microphone was also unable to pickup background noise, thus restricting the noise level in the circuitry.

Historically researchers have advised people to use an external microphone when recording for EVP. This is primarily because, in cassette recorders, it has been necessary to isolate the microphone from the drive motor. You might see a paradox in this since we also suggest that, if you are using a cassette recorder, you should add background noise so that it may be used by the communicating entity to form EVP.

The drive motors of cassette recorders are noisy and this noise is in a range that is not particularly useful for EVP. The motor noise tends to drown out some of the harder to hear EVP.

IC recorders do not always have a jack for an external microphone. They also do not have a drive motor. The electrical circuit in IC recorders is relatively noisy compared to the typical cassette recorder, but unlike the motor noise in a tape recorder, the audio noise generated by IC recorder circuitry has proven to be ideal for EVP. For this reason, an external microphone is not necessary on an IC recorder.

Audio Tape: Selecting recording tape is a matter of what you expect to do with the record of your experiment. If you intend to experiment with the intention of establishing some form of proof as to the authenticity of EVP, then use a new audio tape and maintain an accurate record concerning what is on the tape and where it is located. It is common practice to record consecutive experiments on the same tape while keeping track of the counter number to identify each session.

The sound created by the audio tape, which is known as "tape hiss," can make hearing Class C EVP more difficult than necessary. The use of low noise, high sensitivity audio tape will help in this regard. Also, if a computer is being used to review the audio tape, then there are a number of software tools on the market that will effectively remove tape hiss.

Keeping audio tapes in an archive can be very cumbersome. Most researchers have computers and now transfer their EVP sessions into the computer for review. If something of interest is found, it can be saved and archived on the computer. It is strongly recommended that sound files are archived on external mass storage media to avoid loss due to hard disc failure in the computer.

Background Sound Source: Since sounds in the environment help our invisible friends create EVP messages, the voices will often be found in the noise. Often, the less audible Class B or C messages will be so hidden in the noise that they can barely be heard or understood. Sound editing software can be used to remove some of the noise if it is the steady-state sound energy generated by a fan or radio static.

Amplitude Modulated (AM) radio static has been shown to be more useful for EVP than Frequency Modulated (FM) radio static. The static varies in both frequency and volume in AM radio while it varies only in frequency in FM radio.

More than one sound source, especially if they are quite different in frequency, will tend to produce simultaneous EVP. That is, you might use the sound of running water, which offers a group of high frequencies, and a fan, which usually offers a group of lower frequencies. Under such circumstances, it is not uncommon to find two distinct EVP at the same point in the sound track, one using the water and one using the fan. It can be very difficult to make sense of either message in this circumstance. So, if you are using more than one sound source try to have them all in the same frequency range.

At present we normally use four to five radios tuned-off station when recording. Several of these are tube-type radios, which we routinely ask the communicating entities to use for direct, real-time responses to our questions. Such communication has been accomplished in the past, and other than the experimenter, the key element in those contacts appears to have been the use of tubes in the equipment, rather than transistors. The radios are set so that the resulting noise from static does not drown out our voice on the recording. There are also two fans in the room that are sometimes used for personal comfort and as added noise. For variety, one radio tuned to a foreign language station, is occasionally used as the only background noise. Throughout all of her years of recording, Sarah Estep has always used what she calls the air band, in the area of the dial between 123 and 129 MHz, as a background sound source. These are the frequency that pilots use to talk to ground control.

Please note that some people believe background sound is not necessary. For instance, many people who only record during hauntings investigations collect excellent EVP. Since they are not adding background sound, these people will often insist that doing so is not necessary. Yet, we do find that there is ample background sound present in these recordings when we review their examples. Also, the recorder of choice for hauntings investigators is turning out to be the IC recorder. These little devices produce considerable internal electrical noise in the soundtrack because of the voice activated feature, signal compression and threshold switching. So in effect, the electrical circuit of an IC recorder supplies the background sound we believe to be necessary for the formation of EVP. In controlled experiments in our home using an IC recorder we have found that using the type of steady-state noise source described above produces more EVP messages as compared to no noise source.

Other Environmental Modifiers: After successfully recording the phenomenal voices, you may want to try experimenting with different techniques. You will find that experimenters try all sorts of devices and energy sources to help the entities communicate, often with good results. Researchers have reported improved results when they added the oddest things. At one time, we along with others in the AA-EVP, were adding an ultrasonic rodent repellant device or the small one that supposedly protects an individual from mosquitoes, to our recording setup. A change in routine often improves results when recording EVP. This may be caused by the increased enthusiasm of the experimenter. Do not be afraid to try different things and especially pay attention to your intuitive or "guided" inclination to try something new. As described in this book, the first EVP we collected was *"Crystals Help."* This was a message of encouragement recorded after Lisa had placed crystals around the recording equipment in response to an inner urging.

You can also "enrich" the sound by changing the acoustical quality of the sound source that you are using. The microphone or IC recorder can be placed in a trumpet, the acoustical chamber of a musical instrument or even between the openings of two water glasses. Your objective is to have sound energy that is rich in frequency, amplitude and the sort of phase variations that occur when sound is reflected from hard surfaces.

EVP on the Reverse Direction of a Sound Track: EVP experimenters long ago discovered that there were EVP voices on the reverse direction of the sound track. You may want to check for this. After you review the sound track for voices in the usual way, listen to it again while playing it backwards. Done properly, your voice will be heard in reverse, and the EVP voices will be heard as if they were being played forward. There is an example recorded by Sarah Estep at aaevp.com. Listening to the reverse direction of the recording is easiest with a computer based sound editor, such as Audition.[18] Reverse playing cassette players are also available, and if you are mechanically inclined, it is possible to "break" an auto-reversing tape player by disabling the mechanism that moves the "read" head. Other than rerecording the sound track on to audio tape, the only method we know to listen to the reverse direction of a sound track recorded with an IC recorder is with a computer. In cassettes, the objective is not to turn the tape around and play Side B, but to actually play Side A backwards.

No, we do not know how it is possible to have forward spoken voices on the sound track when it is played in reverse. This is one of the characteristics of EVP that tends to discredit many theories designed to explain these phenomena. The concept that all time in the Physical Plane is "now," for an entity existing outside of the Physical Plane, might explain this. There may also be an explanation hidden away in the idea that time is nonlocal, and therefore holographic-like, because it is an aspect of the operation of reality, which we think of as Natural Law. Speculation abounds.

Computer Recording: A computer may be substituted for the tape recorder. The computer should have an audio input jack, speakers, headphone jack and sound recorder/player application of some form. Microsoft Windows comes with a Sound Recorder application that will work. A sound editor like Audition[18] is most popular as a computer based tape recorder, because these applications allow for easy amplification, filtering and reversing of the sound files. At this time, we do not know of a similar program for Macintosh users, although Mac users often resort to using the sound editing capability of video editing software.

An EVP experimental session can be recorded with an audio recorder and then transferred into a computer for review, editing and storage, or a microphone can be attached directly to a computer and a sound editor can be used as the audio recorder. There are instructions in Appendix B and at aaevp.com that will help you set up your computer for recording and for transferring sound tracks from an external recorder to a computer.

Analysis of the Recording: Lisa is the one in our family who is most skilled at listening to sound tracks and finding EVP. She has been doing this for over fourteen years, and her sense of hearing and her sense of what are and are not EVP is highly attuned. Until the more powerful computers were available, we used a component-type reel-to-reel or cassette tape recorder/player to listen to sound tracks. This process necessitated our using the counter on the tape deck to locate places on the tape that had EVP. It also required that we were able to "rock" the tape back and forth over a specific place on the tape in order to repeatedly listen to the same segment of tape. In fact, we often listened to the same bit of tape dozens of time in an attempt to understand a possible Class B EVP. For this reason, we always insisted on using a tape deck that had control buttons that were situated like piano keys for easy access and that were mechanically linked to the tape transport mechanism to assure positive response to start and stop commands.

Today, we use an IC recorder, download the audio files to the computer and use Audition to listen to EVP recordings. This transition did require a learning curve as we learned how to transfer the sound track from the tape recorder to the computer and how to use Audition. However, once mastered, this new procedure has proven to be a tremendous improvement over the previous method.

One of the benefits of using a computer-based system to look for EVP on a sound track is that the sound editor can be put into "Loop" mode, allowing the segment of sound to be played over and over again. Also, the entire track can be reversed with a single command and then returned to normal with a second command. All of the capabilities of a tape deck are available in sound editing software, and are easier to use.

The constant complaint of early researchers was that phenomenal messages were hidden in the noise. This problem has been solved to some extent by the introduction of sound editing software. It is now possible to separate a voice that is deeply hidden by background noise so that it may be clearly heard. Otherwise, the message would be lost if just a tape player were used.

Unless there is some compelling reason, we no longer take a lot of time trying to decipher EVP that are not immediately understandable. When we first started recording, we took the time because we wanted to understand every word. It also took us a while to develop a communication bridge to bring in contacts that produced Class A voices. The voices are often whispers in the early recordings made by a new experimenter, and sound editing software makes it possible to amplify sections of the sound track in which the voices may appear. Many researchers say that experimenters should not spend time with Class C EVP at all. They reason that it takes too much of the experimenters time, and that if the entities know that they must produce Class A and B EVP in order for the experimenter to hear them, they will do so.

Caution should be exercised in over amplifying a Class C message. There is the possibility that a certain amount of energy from radio or television broadcasts might find its way into your recording equipment. Such bleed over is usually obvious, but when you amplify a very weak EVP until it is easily understood, you may be amplifying a radio broadcast as well. Use discernment when accepting EVP that have been enhanced in this way.

As a second caution, the meaning of very weak EVP can be changed, depending on the way it is enhanced and the sequence in which the enhancing steps are applied. At some point, it is best to give up on a possible EVP, rather than risk interpreting it incorrectly.

A step-by-step procedure for performing some editing functions is offered at aaevp.com and also in Appendix B.

Something to Remember about Editing Your EVP Samples: Some "witnesses," people who are not experimenters but who listen to EVP samples and pass judgment as to whether or not EVP is a nonphysical phenomena, will discount an EVP without further discussion if the sample is modified in any way. If extra background sound is present, if the experimenter's voice is not in the recording, even if the sound track has only been amplified, some witnesses argue that the sample is disqualified from consideration. Granted, this is an extreme attitude, but it exists and you should be aware that you may encounter such rejection if you choose to present your EVP to the public.

Our position is that there are two different purposes for EVP. First, most people seek to use EVP as a means of making contact with a discarnate loved one or an entity in a hauntings situation. For these people, amplifying an EVP to make the voice understandable is acceptable. Second, some people use EVP as a research tool, either as an aid for the study of other fields of interest or as a means of proving the existence of EVP and understanding the nature of its reality. For these investigators, an absolute control over recording conditions and treatment of the sound track is essential.

Building a Bridge Across the Veil: Some of the best EVP are collected by experimenters when they are enthusiastic about trying a new technique or a new piece of hardware. It has been demonstrated that people who have a strong motivation to collect EVP are more successful than people who have come to look upon their recording sessions as routine, and maybe even obligatory. Try to think of each recording session as a new challenge. Enthusiasm and other energetic attitudes apparently provide a special form of energy that communicating entities can readily use to bring information into the physical.

As we have previously discussed, EVP experimentation is a process of building a bridge between our world and the world of the communicating entity. If logic can be applied here, just as you must first decide to communicate and then make the effort to do so by setting up equipment and setting aside the time to experiment, so must the communicating entities decide to work with you and make the necessary arrangements to "meet" when you turn on the recording equipment.

If it is your intention to establish a communication link with a loved one or an entity who may be able to help you in some way, you are in effect, asking a person to take the time to work with you. This is very much as asking a living person to take the time to meet. Many EVP experimenters have come to believe that the communicating entity is the one who actually selects the experimenter they will work with.

Collecting EVP in the controlled conditions of your home is probably best thought of as a cooperative effort. Many experimenters develop a relationship with a spirit guide on the other side who continues to work with them on receiving EVP and who helps bring to them those with whom they would like to communicate. When you first begin your EVP experiments, we would suggest that you ask for just such a spirit guide to assist you.

Location: There is some evidence that it is possible to build up a field of energy in a specific location, which is beneficial for passing information across the veil. You may be familiar with the use of a "cabinet" in séances for physical phenomena. The physical medium sits in an enclosed area and the other sitters in the séance sit outside of the enclosure in a semicircle. The enclosure acts as a container for the psi energy of the physical medium and the sitters, and the accumulation of energy in the enclosure is then used by the communicating entities to produce phenomena.

Group energy such as this is not always thought of as being contained in an enclosure, but may also accumulate in a virtual container defined by the network of attention represented by like-minded people. It is entirely possible that the international network of EVP and ITC researchers represent a contact field that facilitates their collection of these phenomenal messages.

Psi energy can be accumulated more quickly by like-minded people working in concert toward a common goal. This form of energy accumulation is sometimes referred to as a "Contact Field" or the energy of "Rapport." Many researchers feel that EVP contact is more easily achieved when researchers work in a strong cohesive group rather than individually.

It is possible to quickly dissipate this energy if "negative" or dissenting thought is injected into the group. For instance, if you have built up a strong contact field in the room you use for EVP experimentation, and then invite a visitor into it who holds considerable doubt in what you are doing, the person's doubt will tend to dissipate the contact field.

A specific room in our home is set aside for EVP and ITC experimentation. Other venues have produced successful experiments for us, but we think of the experiment room as a battery of sorts for all of our experiments. The equipment that we use for experiments is kept separate from the equipment used for mundane pursuits. If someone wishes to visit us to witness an experiment, we are very careful to "sense" his or her agreement in what we are doing. This is not to say that we exclude anyone who is not already "a true believer" in EVP. It is the people who see themselves as living in a negative world, or who tend to find the downside of just about everything around them, that we would rather not have in our home and certainly not in the experiment room.

The concept of rapport, or a contact field, represents a subtle influence that is a measurable factor in your success experimenting with these phenomena. You will find that avoiding exposing your environment to undue negative influences, while seeking cooperation with like-minded people, will have a beneficial influence in other areas of your life as well.

Chapter 11

What We Know About EVP

Think of EVP as a form of direct communication with people who are no longer living in the physical world. Information has been received from entities through EVP and ITC that show we can also receive messages from those who have never lived on earth. These entities state that they live or have lived in dimensions other than our aspect of reality. Also, messages have been received from those we would consider as extraterrestrials who are very much alive in our reality. Information has even come through EVP and ITC that points to the source as being that of devic energy, which includes nature spirits and other entities who may never have experienced a lifetime as a human. However, we will explain in this chapter that we have good reasons to believe that most paranormal voices come from those who once lived in the physical and who now live in another realm of existence.

The best thing about EVP is that you do not have to take our word that the paranormal voices can be recorded. You can prove it to yourself by simply conducting EVP experiments for yourself. As you learn to record EVP, you will most likely want to tell your family and friends about these exciting phenomena. What you will find is that many people will simply not believe you. EVP is way past most people's limit of acceptance because it simply does not fit into their worldview. Keeping this in mind, we believe that it is best to drop the subject of EVP and ITC with anyone who reacts in a negative way. The alternative seems to be to evoke a potentially hostile reaction from people as they attempt to preserve their sense of what is real in the world.

Interestingly, some religions embrace EVP and others think it is the work of the Devil. It might be a surprise as to which is which. If information which has come from the Vatican is any indication, some leaders of the Catholic religion have been supportive of research in EVP.[29] Of course, Spiritualists and Spiritists find many points of agreement between EVP and their concepts. One of the points of agreement between people who study EVP and ITC and Spiritualists is that both are concerned with the need for a solid foundation for anything they believe. Most researchers who study EVP do all they can to apply good science to their experiments. In the same way, Spiritualists take the "Science" part of Spiritualism seriously and think of EVP as "science-based" phenomena.

Characteristics of EVP

By definition, EVP are unexpected voices that are collected onto digital and analog recording medium and that are not explained by currently known physical principles. They appear to be ubiquitous in that experimenters around the world are able to collect them with just about anything that will record human voice frequencies and under just about any recording circumstance. They often vary in nature as the experimenter, recording environment and technique is changed. English speaking people tend to collect English EVP phrases. Multilingual people often collect EVP in a mixture of the languages that they speak. People who have a strong music background often record phrases that are sung, rather than spoken.

The following list of characteristics will provide a sense of how EVP sounds and the nature of the phenomenal voices.

1. **EVP are Distinctive**: EVP have a distinctive character of cadence, pitch, frequency, volume and use of background sound. The voices have a distinctive sound to them that is difficult to describe. For instance, EVP messages often have an unusual speed of enunciation; the words seem to be spoken more quickly than normal human speech. Regarding this peculiarity, Konstantinos[40] wrote, "The best way I can describe it is that it's almost as if each word is spoken quickly, yet the pauses between the words are of a natural length. The combination of these two speed factors makes for the peculiar rhythm and perceived speed." You may also notice that the paranormal voices often have a hollow and/or monotone quality.

2. **Frequency Range**: EVP are sometimes received at higher or lower time-base than normal speech. The enunciation of words is

not just faster, but the frequency range of the phrases is some-times higher than normal human speech. Such recordings are typically adjusted up to plus or minus ten percent of the initial time-base (sped up or slowed down) during analysis to make them more understandable.

3. **Missing Frequencies**: Italian researcher, Paolo Presi,[31] has re-ported that spectral analysis of EVP samples has shown that the fundamental frequencies of voice associated with the human voice box are sometimes missing in EVP. He describes the typi-cal EVP as a "thickening" of the background noise to form the voice. It is noted by others that the wave form of an EVP tends to have a relatively slow "rise" and "fall" time, when compared to normal human speech.

4. **Precursor Sounds**: Sounds are often heard prior to an occur-rence of EVP. Although these vary in nature, they tend to be within tenths of a second of a phrase and are a "popping" or "clicking" noise reminiscent of the "squelch" sound caused when the automatic gain control engages as the "push to talk" button is depressed on a Citizen's Band radio. There are various ideas on what this sound may signify. The sound may be caused by a di-mensional breakthrough and may be an artifact of spirit world energy entering the physical world. There is evidence that "Psy-chic Time" flows differently than "Physical Time" and so the noise may be caused by a shift in time as the two aspects of real-ity link up. Many researchers feel that the sound suggests that the communicating entity is using some form of mechanism to assist in communications.

As Dr. Ernst Senkowski described this aspect of EVP,[54(V9N2)] "Several observations clearly show 'carrier' signals apparently following a remarkable reduction of the volume/noise coming from the radio receiver and starting with a sharp click like switching on. Afterwards, the receivers may be 'dead' for some time. So, in our system and with our words, we suppose someone transmitting signals which supermodulate our electromagnetic-acoustic field. In my view, this is a clear sign for a special 'field' not yet to be measured by our instruments but overlaying and manipulating our space-time 'reality' which is tightly bound to certain functions of our mind-brain system."

5. **EVP Show Evidence of Being Limited by Available Energy**: Alexander MacRae[30] has noted that the utterances tend to have about the same amount of audio power in their associated sound wave from one EVP sample to another. That is, a short EVP will tend to be louder than a long EVP. A very long phrase might be composed of two or more average length phrases separated by minor pauses. Again, this is as if a communicating entity is attempting to manage available power as "packets" of sound or psi energy. The evidence is very strong that EVP are energy-limited phenomena. This question will be addressed later in this chapter.

6. **EVP are Complete Words or Phrases**: Researcher Alexander MacRae, has also conducted considerable analysis of EVP messages, determining that a message is typically one to two seconds in duration and is not truncated at the beginning or end. If EVP were crosstalk, they would often begin in the middle of a word. EVP messages are usually complete thoughts, as well.

7. **EVP are in the Language of the Experimenter**: Alexander MacRae has conducted experiments in a place that has no English language radio or television stations, yet resulting EVP were in English, which is his primary language. It is typical for the EVP, no matter where they are recorded, to be in a language that the experimenter understands. There have been exceptions to this, but as a rule, experimenters will hear EVP in their native language, or a language in which they are conversant.

This brings up an interesting point of speculation about psi-based communication. Mental mediums often report that they receive communication from nonphysical entities as images which they must interpret. These images are not just mental pictures. They are packets of information that are sufficiently complete for the receiver to fully understand their meaning. Robert Monroe[4] referred to this form of information as a, "Thought Balls."

There are numerous reports of EVP that are apparently initiated by extraterrestrials. It seems safe to assume that extraterrestrials do not speak English. Is it possible that, in EVP, we have proof of the nature of telepathy? Are images the standard mode of communication in nonphysical aspects of reality? If so, our Self/brain mechanism must be an effective translator of images into recognizable words in a language we understand.

8. **EVP are not Ambient Sound or Broadcast Programming**: Again, Alexander MacRae has made a contribution to the field of EVP by submitting the newest model of his Alpha Device for testing at the Institute of Noetic Sciences. The device produced EVP in a chamber that was shielded from both Radio Frequency (RF) and sound energy in the environment. This demonstration proving that EVP are not stray sound or RF has been made before. One of the problems we have faced in the past is that "old proof" is often discounted because of the considerable improvement in instrumentation, experimental protocol and understanding of physical principles. In light of this, it is important that our modern generation of scientists is able to witness demonstrations that adhere to modern standards for research, such as that just provided by Alexander MacRae. As of the writing of this book, we await a response from the scientists.

9. **EVP are Appropriate to the Circumstances**: There are numerous examples of EVP that are clearly direct responses to questions recorded just prior to the EVP phrase or to the circumstances. As an example of an EVP being appropriate to a circumstance, we were trying to figure out how to set up a new tape recorder while we were preparing to conduct an experiment. The tape recorder was finally set up correctly but not before causing a very loud feedback squeal that was recorded. On the recording a male voice said in a Class A EVP, *"Leave it alone"* right after the loud squeal. Considering the circumstances, we took that to be good advice, as tinkering with the equipment had caused the noise and we had finally set it up correctly.

EVP can be obscure in meaning in that, without a reference point as to the subject, the message may sound meaningful yet may not make sense to the listener. If we had not recorded the squeal, therefore knowing what had just taken place, the EVP *"Leave it alone"* would have had little meaning for us.

10. **Precognitive Responses**: Interestingly, answers to questions may be recorded prior to the question being asked, so that the an-

swer as a phenomenal message is on the sound track followed by the experimenter asking the question. While time may be meaningful to us, our time may well be irrelevant to a nonphysical entity. Alternatively, the entity may be sensing what the experimenter is about to ask. The transcript of a telephone call in Chapter 8, which was provided by Sonia Rinaldi, includes several instances in which the entity answered a question before or while it was being asked.

11. **EVP are Found by Playing the Soundtrack Backwards**: One of the more bizarre characteristics of EVP is that it is possible to discover an EVP that seems to be garbled, but that makes perfect sense when the soundtrack is played in reverse. By this, we mean to say that the sound track is played so that the voice of the experimenter can be heard speaking backwards, but the EVP can be heard speaking forwards.

    As with the ability of communicating entities to anticipate questions by placing answers into recording media before the question is asked, the phenomenon of reverse track EVP provides important hints as to the nature of time.

12. **Vocalized Questions Elicit more EVP**: There is evidence that the communicating entities are able to read our thoughts, as in placing an answer on a recording prior to our asking the question. However, experiments conducted by Alexander MacRae[30] have shown that EVP responses increase when questions are asked out loud. MacRae conducted a simple experiment during which he ran numerous sessions and did not verbalize questions, and then the same number of sessions speaking the questions. He then counted the number of EVP responses. On the non-verbalized sessions he collected 3.2 utterances per session, while as on the verbalized sessions he collected 5.3 utterances per session.

13. **The Voices in EVP are often Recognizable**: If you have ever felt that EVP experimenters are mistaking stray radio broadcasts for phenomena, you should know that it is common for an EVP to contain the recognizable voice of a deceased person. It is also common for that entity to say something that was typical of what they would say when in the physical. Their personality clearly

remains intact even though they no longer have a physical body. There are just too many reports of a loved one's voice being recognizable to ignore this fact, even after they have been on the other side for many years.

14. **Mundane Voices are Sometimes Transfigured**: A communicating entity will sometimes remodulate or transfigure the experimenter's words into EVP. In one striking example, the words of a French speaking radio announcer were changed, mid sentence, into an English spoken EVP. The EVP was clearly inappropriate for what the announcer had been saying. By the way, here we use the term "transfigured" in much the same way that it is used in mediumship to describe how an entity transfigures or changes the medium's features into the entity's likeness. Many successful experimenters use a foreign language radio station or recording for background noise and have great success in receiving loud message in their own language.

15. **Party Line**: Some EVP sound as if they are comments intended for someone other than the experimenter. This is much like momentarily listening in on a party line telephone call. It is not uncommon in both field and controlled recording situations to record comments that seem as if unseen people are discussing the experimenter's actions in much the same way that you might discuss the activity of someone that you were watching.

16. **A Need for Background Sound Sources**: As has been noted, sound energy is often provided during EVP experiments in controlled situations. It is felt that the internal noise in IC recorders is used by the communicating entities to form EVP in lieu of supplied background sound. The conclusion is that communicating entities require such sound energy as raw material to generate EVP. As is noted in Item 4, there is also evidence that background sound energy is "collected" prior to an EVP, as if the words are being "burst" transmitted after sufficient energy has been accumulated.[30] In the same way, EVP may trail off or become garbled near the end, as if available energy is being rapidly depleted.[32]

17. **Recording Media Stability**: It has been reported that EVP messages stored on audio tape will sometimes change over time. For

instance, an experimenter notes the presence of an EVP in his or her log book, but on later examination of the recording find the message changed or no longer present. So far we have not heard that EVP recorded on compact disc has changed. Some researchers have said that audio tape will degrade over time and should be re-recorded after several years. Interestingly, copying an audiotape is apt to contaminate it with new EVP.

18. **Layered EVP**: Experimenters often complain about having several layers of EVP in the same location of the recording media. This is especially common when more than one background sound source is used during the experiment. For instance, if a fan and radio static is used for background sound, a message might be found in the fan noise and a second message might be found in the radio static—both in the same segment of sound track. This is one of the reasons we say that there is something of a "learning curve" for learning how to hear Class C and Class B EVP. To detect every last EVP, you must learn to selectively listen to each contributing sound on the track if you wish.

19. **The "Newness" Effect**: The experimenter's excitement in trying a new detection device or recording technique may be the source of improved EVP collection. As the new approach becomes "normal operating procedure," the improvements generally fade back to a more "normal" frequency and quality of EVP collection. This suggests that it is important for the experimenter to maintain peaked interest during experiments. This is also one of the reasons we believe that the experimenter is an integral part of the recording circuit. The experimenter is clearly supplying the necessary psi energy to enable a nonphysical to physical transfer of energy.

20. **Effective Devices Unique to the Experimenter**: Exceptionally effective EVP and ITC collecting systems have been developed; however, these typically work well for the developer, but do not work as well for other experimenters.[31] This paradox supports the belief that the experimenter is part of the recording circuit. It has also reinforced the concept that the communicating entity may be specific to the experimenter.

Because various devices and equipment setups have worked exceptionally well for one experimenter and not for another, it has been argued that it is a waste of time to try to develop the hoped for equipment that will become the proverbial "spirit telephone" that anyone would be able to use to call up his or her loved one on the other side. This may prove to be the wrong assumption. There is growing evidence that people who had not been able to record EVP with a cassette recorder are now doing so on IC recorders, and people who had recorded mostly class C and B EVP are now collecting many Class A voices using this type of recorder. This argues that technological improvements are possible for EVP.

It is reasonable to expect that the thousands of hours spent by EVP researchers around the world will result in new devices and techniques. Such advances, teamed with our growing understanding of these phenomena, should assure that better and more meaningful EVP and ITC collection will be possible in the future. Nevertheless, the attitude of the experimenter seems to be the dominant success factor and the quality of a particular device is less important than how and by whom the device is used.

21. **EVP can be Thoughts of Living People**: There have been a number of well designed experiments that appear to have resulted in EVP initiated by living people who were sleeping at the time. As an ethical consideration, such experiments are always prearranged with the person who volunteers to be the sleeping target. In these experiments, questions are clearly answered by a communicating entity, and the answers are appropriate for the sleeping person. This fact of EVP suggests the possibility that EVP can become an important tool for consciousness research. For instance, is it possible that a patient in a coma might initiate an EVP when requested? This fact also offers support for the Survival Hypothesis, in that it demonstrates that when Self is disentangled from the physical body, as in sleep or meditation, it is very much the same as a discarnate entity. This is a prediction of the Survival Hypothesis.

Do not take the fact that living people can initiate an EVP to be the explanation for all EVP. The fact that a living person can initiate an EVP is something that must be expected if we are Self in a physical body. When we physically "die," we simply become discarnate Self in an etheric body. When taken in the context of the other evidence about EVP and the Survival Hypothesis, the fact that we can record thoughts of living people provides a substantiation of that other evidence.

Identified EVP characteristics are much more extensive than those listed here, but these are the more noteworthy and most commonly experienced. These characteristics have been noted for years by experimenters and researchers around the world. There are still many questions and much that we do not know about EVP. However, through the continued work of dedicated researchers like Alexander MacRae, Paolo Presi and others we are certain that the list of what we know will continue to expand and grow. EVP as a tool for spirit communication is still evolving. You might think of the current state of EVP evolution as being similar to that of the radio about the time that Heinrich Hertz was experimenting with electromagnetic waves in 1894.

Theories Explaining EVP

There have been many theories offered to explain EVP. To date, all of these theories have failed to provide a complete model that answers all of the evidence except the Survival Hypothesis. Of the theories to explain EVP that you will encounter, the technology artifacts, normal physiological response, Survival Hypothesis and Quantum Holographic Universe Hypothesis are dominant. Each of these will be briefly discussed with an attempt to relate them to the evidence.

The Technology Artifacts Argument: The "artifact" argument holds that EVP are the result of radio or television broadcasts bleeding into the electronic circuit or are natural manifestations of electronic circuitry.

It is possible to record the signal from a local radio station in an audio recorder. This is not common, but we are aware of instances in which people have used IC recorders for recording EVP in a hauntings investigation, in which mundane voices were picked up from a radio transmitter that was located near the site. EVP experimenters should always be aware of this possibility. Again, this is rare and an experienced experimenter will be able to distinguish between a radio broadcast and an EVP.

It is also possible to mistake a burst of noise caused by something like a bumped microphone for an EVP. As will be discussed under "normal physiological responses," it is not uncommon for a person to find words in what is actually just random noise. However, and this is an important "however," random noise is audio energy and communicating entities will use such energy to form words.

One of the characteristics of the Panasonic RR-DR60 IC Recorder, which is one of the earliest models of the recorders often used for field recording, is that the internal circuitry will tend to cause loud bursts of noise that sometimes lasts about as long as it takes to speak a word. The bursts of noise are probably true artifacts of the recorder, but as it turns out, the noise is often used by an entity to form words. Researchers still know little about the nature of EVP and how EVP are formed. Thus, we do not know if the communicating entity is causing a perturbation in the circuitry that will cause a noise so that it can speak, or if the entity is simply taking advantage of burst of noise within the circuitry to speak.

The "artifact" and "broadcast" argument are usually the first to be voiced by skeptics who are new to the field. These were also the first possible explanations for EVP that were tested by past and present EVP researchers. These possibilities have been eliminated by researchers working with two separate audio recorders, or as Alexander MacRae did by recording in a compartment that is screened from radio frequency and sound energy that is in the environment. These experiments have been conducted numerous times. EVP will occur on one of the two recorders while ambient (environmental) sounds will occur on both, thereby eliminating the "artifact" hypothesis. EVP will occur in a screened compartment even though the compartment is effectively shielded from all possible types of radio waves, laser beams, electromagnetic radiation and all acoustic energy, including audible sound, infra-sound and ultrasonic waves, thereby eliminating the "broadcast" hypothesis.

The Normal Physiological Responses Argument: The argument is that EVP are actually a normal physiological response to mundane noise. In this, it is maintained that the mind is trained to recognize patterns in noise, such as human voices or human faces. Thus, EVP are random noises that the mind forces into recognizable words, even though such words are not present.

Again, we will concede that this does occur. However, this does not account for instances in which the average person will agree on what is said in a Class A EVP, even without prompting.

In the June 1974 issue of the *Journal of the Society for Psychical Research,*[61] "Letters to the Editor," renowned psi researcher, Scott Rogo, wrote that, "I want to point out a fallacy that critics have made about tape recorded voices. It is claimed that, because several individuals have different interpretations of a 'voice,' that disqualifies that voice. Instead, it is some mechanical or accidental sound that really exists on a tape. This is based on the idea that if a voice did occur on the tape, all listeners would interpret it uniformly."

Scott goes on to show the fallacy of this premise and mentions experiments by Dr. John Lily, known for his work with dolphins. In this case, Lily presented a tape of a clearly enunciated word spoken by a human voice. It was discovered that individuals heard up to thirty different words, for this one word. Scott concluded his letter with, "This seems to rule out the argument that the lack of unanimity among the listeners must destroy confidence in tape recorded voices."

It has been shown that any audio message that is very low in level, compared to the background noise, is difficult to understand. Following the work of Paolo Presi[31] and his team, some EVP do not have the typical spread of frequencies and are often abnormally high or low pitched. Much of the difficulty people have agreeing on what a Class C or Class B EVP is saying can be attributed to the absence of some of the more important cues we depend on to interpret speech.

In the end, this explanation for EVP can be discounted because of the numerous listening panel tests that have shown selected EVP to be clearly understood and meaningful in content.

Quantum Holographic Universe Hypothesis: The Quantum Holographic Universe (QHU) Hypothesis argues that there is a field of energy underlying the universe of physical matter that has not yet been quantified. Further, that information never ceases to exist. It is nonlocal in that any part of the universe contains all of the information there is. Information is accessible through the human senses.

This hypothesis is often used to explain many observed nonphysical phenomena as "echoes of the past." Edgar Mitchell[24] is a strong proponent of a variation of this hypothesis that includes the concept of "spontaneous" occurrences of life and/or intelligence, thus eliminating the need for a creator entity. You can also find considerable information about variations of this hypothesis in the book, *The Holographic Universe* by Michael Talbot.[33]

A holographic image of an object can be made by splitting a beam of coherent light, which is laser light, and shining the resulting two beams onto a photographic plate after reflecting one beam off of that object. The image produced on the photographic plate is an array of dots representing an interference pattern that is caused by the combined effect of the different distances the two beams travel and the information caused by the beam's reflection from the target object. The image on the photographic plate can then be projected as a three-dimensional rendition of the target object by essentially reversing the projection path of the laser. You can use a very small portion of the photographic plate to reproduce the whole object; however, an image from a portion of the plate will not be as clear as that produced using the entire plate.

The fact that a portion of the photographic plate contains essentially all of the information necessary to reproduce the image is referred to as nonlocality, meaning that all of the information can be found in any part of the whole. Since it seems possible for a person to psychically access information that should only exist in one part of the universe, no matter where the person is, it is a natural leap in logic to argue that such information is nonlocal—just like the information on a holographic plate.

The QHU Hypothesis acknowledges the existence of a portion of reality that is not yet defined by principles of physical science. However, this uncharted aspect of reality is not considered to be "nonphysical" in nature. This is not a subtle difference. Metaphysical cosmologies usually describe reality as consisting of planes or levels of existence. In the QHU Hypothesis, there is but one level of existence and there is no such thing as a Self existing in other planes of existence.

As will be discussed later in the chapters on Video ITC, we have noted holographic-like characteristics in Video ITC as well. In fact, there is growing evidence that the QHU Hypothesis may correctly explain certain aspects of reality. For instance, the existence of EVP found by playing a sound track backwards, the multitude of faces found in Video ITC and the accessibility of nonphysical information by anyone living in any part of the world, seems to be best explained using the holographic principle of nonlocality. However, we do not feel that nonlocality accounts for the existence of a nonphysical aspect of Self. This is an important omission because we have substantial evidence that we, as Self, survive physical death and that we retain our personality.

The QHU Hypothesis does not account for the real-time intelligence that is obviously initiating contact. It does not allow for entities answering direct questions, which is seen in many EVP messages. For instance, one open-minded skeptic within the AA-EVP asked tapers to ask communicating entities to record the words "Mary had a little lamb." One experimenter finally recorded the words sufficiently clear for the skeptic to hear them. The experimenter asked the communicating entity to do something specific. If EVP were a product of a Holographic Universe, such a direct response would not be possible. In addition, as is shown in the chapters containing examples of EVP, the communicating entities show that they are very much aware of us and often comment on what we are wearing or something that is different in the recording area. How could such responses come from a recording of the past? EVP are very much two way communications that often have logical and meaningful responses to our questions and requests. This is nothing like picking up echoes of the past and rules out the QHU Hypothesis.

Survival Hypothesis: It is held in the Survival Hypothesis that the real you, the point or perspective from which you experience reality, is Self. Self, is nonphysical in nature and is in a symbiotic relationship with a physical body during a lifetime. When the physical body dies, Self (its attention) returns to nonphysical reality. It is the nonphysical aspect of Self that initiates EVP.

The Survival Hypothesis must be taken in the context of a metaphysical cosmology to make sense. The existence of a nonphysical aspect of Self existing in a reality as described by metaphysical concepts does account for most of the characteristics of both EVP and

Video ITC. For instance, the responsiveness of the EVP originators to questions and recording circumstances cannot be explained as "echoes of the past" as argued in the QHU Hypothesis, but requires the existence of an unseen intelligence.

The body of evidence gathered through past revelations, which have been brought to us through various forms of mediumship, tends to agree with information that is collected through EVP. Communicating entities sometimes describe their environment in ways that are consistent with known metaphysical cosmologies. They also sometimes indicate that they must "go on" to aspects of reality that are "out of range" for EVP communication. This latter point seems to argue that there is a hierarchy of levels of existence in reality.

The AA-EVP NewsJournal has the slogan, "Founded in 1982 by Sarah Estep to Provide Objective Evidence That We Survive Death in an Individual Conscious State." The "survival" phrase had been in the NewsJournals published by Sarah and we decided to continue that tradition when we assumed leadership of the Association, because the evidence received through EVP and ITC is overwhelmingly supportive of the Survival Hypothesis. It is immensely evident that there is a greater reality of which our physical aspect is just a small part, and that we exist in that larger reality. Our experience in this lifetime is both a temporary one and one that is only dominant in our awareness because that is where our point of view is focused during physical life. Our natural state is nonphysical in nature.

The Survival Hypothesis and some of the Quantum Holographic Universe Hypothesis can be complementary. The Survival Hypothesis helps to explain the "who," "why" and some of the "how," while the QHU Hypothesis may help to explain some of the "how." For the "who," we know that discarnate entities are amongst those who are initiating our EVP messages. For the "why," we know that these entities are as eager to communicate with us as we are with them and that they are for many of the same reasons. There is also some evidence that EVP are initiated in an effort to offer guidance that might lead the experimenter to a greater understanding of themselves and the greater reality. This, we believe, is important to assure that the experimenter will know what to do when he or she transitions.

As for the "how," we think that the messages are created as a mind-to-mind transfer of information between the nonphysical entity and the nonphysical aspect of ourselves. That is mediumship and is supported in the Survival Hypothesis. What is not clear is how the nonphysical entity, experimenter and recording apparatus combine to produce the phenomenal messages. The Survival Hypothesis does not address this question to any extent. If reality has a holographic aspect to its operation, then the QHU Hypothesis may help us better understand the "how." For instance, as explained in Chapter 12, stochastic resonance, in combination with the subtle psi energy of telekinesis, may be the mechanism by which the phenomenal information is impressed into our recording media. If this is correct, then there is a strong link between how nonphysical reality behaves and the quantum and holographic principles we are learning to use to explain how our physical aspect of reality operates.

For us, the objective of future research is to find a model that will account for all of the observed characteristics of these phenomena, and we expect that it will be much more complex than what has been noted here.

> "If you wish to upset the law that all crows are black,
> you mustn't seek to show that no crows are;
> it is enough if you prove one single crow to be white."
>
> William James

This quote from William James has become something of a battle cry for people who would prove to scientists that survival is real and that phenomena like EVP and mediumship prove this to be the case. With EVP and mediumship we feel that we actually have a whole flock of white crows. However, you must be the final judge of this. Final proof of anything is always gained through personal experience. Pick up an audio recorder and bring a generous supply of patience as you seek to collect EVP examples for yourself. You can do this, and when you do, we are confident that you will then harbor a white crow of your own.

Chapter 12

What We Know About Video ITC

Figure 12-1:
Standing Man

This chapter is devoted to explaining what we know about Video ITC, what we think we know, and what others have said about the subject. We attempt to address the question of whether or not Video ITC features are true phenomena, as opposed to artifacts of the technology or simply the overactive imagination of the experimenter.

Some of our observations concerning the characteristics of Video ITC features will be discussed. We believe that it is important for you to understand that this aspect of ITC is rich in "hints" about the nature of reality. Also, the very fact that there are observable patterns to the way these features appear offers a form of validation, showing that Video ITC is not a limited occurrence experienced by just a few people under very controlled conditions. It is a phenomenon that can be replicated under many different circumstances, and is, in fact, replicated around the world.

A System of Nomenclature

We have adopted a somewhat specialized vocabulary to describe Video ITC images:

Frame = One frame of a video clip. In the National Television System Committee standard for video used in camcorders sold in the USA, there are about twenty-nine frames per second.

Feature = A recognizable cluster of optical noise, such as a face, that is found in a frame. This is the goal in Video ITC.

Large Feature = A feature that dominates a frame in size. These are often suggestive of human or other forms. A large feature may be a scene. Several of the images that Erland Babcock has captured show nature scenes that appear to have come from another dimension. They cover the full frame.

Inserted Feature (Insert) = A feature that is better defined than other features in the frame. These often look like human or other recognizable forms. The feature often stands out so sharply from the surrounding video noise that it appears to have been "pasted" or deliberately inserted into the frame. Occasionally, we have seen rectangular areas of slightly different noise intensity around these features.

Holographic Feature (face fractal) = A feature that is suggestive of human or other forms but that is not as well defined as an insert and is accompanied by numerous other face fractals.

Since these are not photographs, in the usual sense, we avoid the use of this word when describing Video ITC features. One of our greatest challenges in bringing these features to the public has been the task of expectation management, as we work to dispel a viewer's tendency to look for photographic quality evidence. *These are not photographs **they are paranormal pictures.***

The Nature of Video ITC

As stated in the chapters on EVP, the fact that these phenomena are complex with many characteristics suggests that they are real and not chance, accidental or illusory. Of course, complexity does not guarantee this in and of itself, but without this complexity it would be difficult to argue their authenticity. It would also be difficult to study these phenomena if they did not occur in many ways, under many different circumstances and if they did not often provide evidence in their content.

The very nature of the way the pictures are received in Video ITC makes them fuzzy so that they are usually not as crisp and clear as real world photographs. The better defined ITC pictures received by several researchers during the 1980s and 90s were downloaded onto their computers by their spirit team and provided much clearer images.

These researchers began their experiments with Video ITC, and then later, their spirit contacts were able to actually access their computers.

Very few people are working with Video ITC at this time. It is our opinion that more people would attempt experiments if they had a step by step method given to them on how to set up their equipment. This in fact was one of the main purposes of this book. When we received our first video ITC image, we were very excited. Later, when we gave Larry Dean instructions on how to set up his equipment, he was successful and it was a thrill for us to hear his excitement when he also found his first image. In the next chapter, we have provided step by step instructions on how to conduct Video ITC experiments with the hope that more people will begin experimenting in this area. The video experiments are a source of excitement and joy for us and we believe they will be for others as well. Each one is unique, and we are always excited to see what images have come through.

Video ITC features are not as clear but are similar to snapshots of people, animals and occasionally scenes. We have received many images. Some we have recognized but the majority of the images are of people or animals that we do not know. As we continue to build a bridge to those working with us on the other side, we hope that the information about what we are seeing will increase. As demonstrated with Figure 9-2, asking a particular person to appear in a Video ITC experimental session can be successful, much like asking for someone during an EVP experiment.

A few of the many images that we have received are posted on the aaevp.com website. As it is today, what we know about Video ITC features is what we surmise from what we see, the information received from our spirit team and what we have observed about the way the phenomena occur. The images are from various eras of history. This can easily be seen from the different periods of clothing that show up. Some images do not appear to be from humans and must be considered extraterrestrial. It has been just a little over a year since we first began working with the images from other worlds. It takes a lot of time to review the experiments and we are pleased with what we have gotten in the very few experiments that we have had time to do. An improvement in the quality of the figures coming through has been seen and we have even received a few images that appear to be moving, as seen on several frames.

Patterns have been observed in the way these phenomena manifest, and from these patterns we have derived a few hypotheses. Experimenters around the world are attempting to test many of these hypotheses, so the list of what is "known" continues to grow. In the following pages, we will discuss a few of the better established hypotheses that are being considered today. But before we begin, it is important to us that you understand that few people are willing to say that any of these hypotheses are the "correct" explanation for EVP or ITC. This is the same as with other types of paranormal physical phenomena like, psychokinesis, direct voice, materialization, apports, levitation and so forth. EVP and ITC are real and occur; however, how they occur is still conjecture.

Video ITC as an Illusion

One of the dominating explanations offered by the scientific community for Video ITC is that humans are genetically predisposed to see faces in noise. In other words, it is argued that humans have been conditioned since life's emergence from that fabled primordial soup to recognize certain shapes in the environment that may be a threat to their survival, food or a potential opportunity for reproduction. Thus, today, it is argued that we will see a face in almost any random arrangement of dots or optical noise, such as what you find in a thicket of bushes, because it is a survival trait to do so.

Probably one of the more commonly cited sources for this argument is an article in the "Neuro Quest" column of the February 2002 *Discover* magazine. That unattributed article cited research suggesting that face recognition was centered in the *fuiform gyrus* regions of the human brain. Based on this finding, it is argued that humans are "hardwired" to recognize faces. Conducting a little research of our own, we have found that probably the more correct interpretation of current research is that, while face recognition does seem to involve that region of the brain, there is little evidence that region is genetically preconditioned to recognize humans alone. Research[23] seems to support the idea that humans must learn to recognize shapes and that it is possible to learn to recognize new shapes as readily as we do faces. This argues that face recognition is not an unavoidable result of looking at optical noise.

The counterpoints to the genetic predisposition argument can take many forms. For instance, if Video ITC features are an illusory product of our mind, we would expect to find prey animals that our mind would see as a threat to our survival. A few cats and dogs, and other animals have been found, but so far, we have found two friendly looking bears, a wolf and a few baby lions. This is hardly a fair representation of animals that might eat us amongst the many features we have collected.

Video ITC as Television Programming

While some people are quick to accept these features as true phenomena, some have argued that we are actually recording stray broadcast programs. That is, that the faces are simply faces of actors in a television program that has somehow found its way into the video loop. On the surface, this seems like a strong argument, but there are a number of factors that make it unlikely that television programming is involved.

Many researchers have conducted Video ITC experiments using a computer monitor or a television set in which the tuner has been removed. They have received images and so the images could not come from stray television signals. Have you received a stray television signal when you have had your computer monitor turned on?

In a typical Video ITC video loopback configuration with the camera just a few inches from the monitor screen, the video camera "sees" only about ten percent of the screen. That is, a twenty inch screen is about sixteen inches by twelve inches or slightly less than two hundred square inches of screen surface. When we first began experimenting, we typically exposed an area that was about four inches by five inches or twenty square inches of screen surface. If you look back to the last time you viewed a television program, the typical human face also takes up about ten percent of the screen, so let us assume that the video camera was pointed directly at a human face on the screen. The face would just about fill the video frame of the camera. Next, we typically focus the camera beyond the television screen so that the rapidly rolling or flashing image (about three cycles per second) which results from the video feedback loop has a "cloudy" texture between the bright and dark extremes. This typically puts the focal plane of the camera some eight to ten inches beyond the television screen. Under these conditions, the human face would be little more than a blob of color, making it very unlikely that our Video ITC features are the result of a stray television program.

There is another factor that leads us to believe we are not recording television programs. The television set is switched to the Video-In circuit, and not to the antenna or cable. Some experimenters even remove the tuner portion of the television set so that the set is incapable of receiving the high-frequency broadcast and converting it into a usable signal for display. This modification of the television receiver has not had an effect on Video ITC collection.

Also, the video loop does some rather drastic things to the image that is displayed on the television screen. The video camera and the television screen have approximately the same number of pixels. When the camera is "looking" at only ten percent of the television screen, it does so with a standard complement of pixels. The result is that ten percent of the television screen pixels are "seen" by one hundred percent of the camera pixels. One hundred percent of the camera pixels are then projected onto one hundred percent of the television screen. The camera then sees only ten percent of that new projection, and if our math is correct, only one percent of the original image. The next time that one percent of the original feature passes through the video camera, the initial image will be reduced to a tenth of a percent and so on at the rate of about three reductions a second. In effect, the video camera/television set combination functions as a microscope to enlarge whatever is on the television screen. Even if we did have images from a television program, they would be quickly enlarged beyond recognition.

Finally, many of the features are found by rotating the video frame in ninety degree increments. If these images were the product of a stray television program, that program would necessarily be featuring an occasional actor who is standing upside down.

So you can see that the video loop does great violence to the video signal and any stray television program would be quickly scrambled into a melee of color, intensity and video pixels.

The Stochastic Resonance Hypothesis

Emerging understanding of how stochastic resonance influences electronic signal integrity provides an interesting area of speculation that may help us explain how these features are formed. The definition for stochastic resonance is: "Noise-controlled onset of order in a complex system." Noise will combine with a weak signal to cause portions of the signal to become stronger, and therefore, detectable. Scientists are looking at this physical phenomenon as a way to recover weak signals that would otherwise be lost in the noise.

The video signal we use for Video ITC is essentially a random signal that is chaotic in nature. Study of chaotic systems has shown that patterns will spontaneously emerge out of the chaotic energy, and that sometimes, these patterns will appear to be recognizable features. Stochastic resonance is apparently the process by which this occurs.

You can see that it would be natural for a person who knows a little about stochastic resonance, and the emergence of order out of chaos, to suspect that the Video ITC examples we present are simply emergent patterns and not faces at all. We acknowledge that, while one would expect an even distribution of color dots and intensity in the video loopback signal we record, we do find splotches of color and intensity. Some of these are even suggestive of recognizable objects, but they are seldom well formed, and finding faces in them is a little like finding faces in clouds. As it turns out, it is these areas of emergent order that are used by the initiator of the ITC features. In fact, these areas of order appear to be a requirement for feature formation.

The images we find in Video ITC are entirely too well detailed to be the result of any random process or our imagination. It is more likely that stochastic resonance is one of the mechanisms by which the communicating entities inject features into the video noise. Remember that the process of stochastic resonance enables a weak signal to produce large effects. Psi energy (thought energy) is a weak signal as it relates to physical processes. Our speculation is that psi energy from the communicating entity is either directly, or as communicated through the experimenter's mediumistic ability, amplified through the process of stochastic resonance in the video noise to form the phenomenal images.

A similar argument has been made by us for EVP and the same idea, in regard to the use of noise for EVP, is noted by Dr. Dean Radin in a memo to AA-EVP member, Karen Camus[54(V21N4)] that, "... noise itself is probably a stimulus for opening the unconscious mind to psychic impressions (although perhaps this includes the conscious mind for highly experienced listeners). The Oracles at Delphi apparently knew about this and used it for thousands of years. The Oracle site itself is located inside long tunnels near the ocean—the result being that not only were the Oracles likely inhaling intoxicating fumes from natural hot springs or vents deep inside the mountain cliffs, but they were also hearing the sounds of the surf echoing and rumbling through the tunnels. That noise (not exactly white noise, more like "red" noise) is said to stimulate audio hallucinations, and in fact this is exactly why to this day we use white noise to stimulate the receiver's imagination during telepathy tests."

Robert Monroe[4] established that changing sound frequency can cause a "frequency following" response in the brain. Knowing this, and knowing that stochastic resonance appears to be the mechanism by which weak psi energy is amplified to cause an effect on chaotic optical energy, we speculate that a similar process might explain why psi functioning is enhanced with the introduction of noise. This reinforces the notion that understanding one aspect of these phenomena might help us understand other aspects.

Scientists are obliged to explain observed phenomena from the perspective of known physical principles. In this view of reality, life's path of evolution, which is sometimes referred to as the "arrow of creation," points from the origins of the physical universe to the present. In this view, life has emerged out of some primordial soup and consciousness is a product of that evolution. Thus a phenomenon must have a mundane explanation based on known physical principles otherwise it simply cannot exist. However, if you add an *arrow of creation* that represents consciousness and that points from outside of our physical aspect of reality, then you have the emergence of all known physical things as described in physical science, but there is also consciousness emerging from somewhere outside of the current scope of physical science. In this view it is possible for consciousness to exist in the physical aspect of reality, but only because of the evolution of the brain. Consciousness, or more specifically, Self, exists outside of the physical and is hosted in the physical by the physical body.

Following this line of logic, Video ITC features are a product of physical processes such as stochastic resonance. But they are only made possible by these processes. They are governed by an external intelligence that takes advantage of these processes.

For the sake of completeness, we will state the Stochastic Resonance Hypothesis here, but do understand that you should use caution in citing this hypothesis as part of a "proof" for EVP or ITC. As of this writing, this hypothesis has not been subjected to rigorous scientific scrutiny. **Stochastic Resonance Hypothesis**: That psi energy is amplified through the process of stochastic resonance acting on chaotic physical energy to form physical phenomena. **Prediction**: ITC features and EVP messages may emerge from any suitably chaotic, physical energy.

The Use of Background Noise in Video ITC

As was pointed out in Chapter 11, it is common for an EVP experimenter to provide some form of background sound during a recording session. Evidence indicates that the communicating entity requires some form of sound energy, which it transfigures into recognizable speech. As the theory goes, the entity's thoughts are nonphysical energy and anything nonphysical must be somehow made to energetically agree with the aspect of reality it will inhabit. This is the Natural Law known as the Principle of Agreement. In practical terms, this is described as embodiment, since we describe the process of coming into agreement as that of taking on a body which is suitable for this level of existence. Our physical body provides the necessary embodiment to permit our Self to energetically agree with the physical aspect of reality. We believe that, in the same way, audio sound provides a means to embody the communicating entity's thoughts.

A testable prediction of the hypothesis that energy must agree with the aspect of reality it will inhabit is that some form of embodiment of nonphysical energy must also occur in Video ITC. In fact, we believe that it is the highly chaotic video loopback energy that is necessary to embody the otherwise nonphysical features we find in Video ITC. Somehow, the communicating entity is able to transfigure the optical energy with the nonphysical information to form the features that we see. There is evidence of a similar requirement for physical energy in many forms of paranormal phenomena.

In our study, we have found that just about any optically noisy environment is liable to contain phenomenal images. We will discuss the variations of this theme later, but here we will state that it is as if nonphysical entities are looking for a way to communicate and are seizing any opportunity to do so. One example is the "Face on the Wall" once found in the Spiritualist Desert Church in Las Vegas,[9] Nevada. The Reverend Catherine Stewart, who recently transitioned, was Pastor at the time. The church has since closed and the face has been lost due to remodeling.

The congregation painted the wall in question shortly after they moved into a new church building. Close examination of the painted wall shows that they may have first used a paint roller and then something like a sponge to daub a medium blue texture onto the light blue wall. A few months after painting the wall, the congregation noticed that they could just make out the outline of a face in the paint. On a visit to the church, we photographed the wall with a digital camera. A little change in the intensity and contrast of the digital picture produced what you see in Figure 12-2.

Figure 12-2: Face on the wall

We have a high confidence that the face was not deliberately painted under the new paint, nor could it have been without leaving telltale level changes in the surface. The feature is also very convincingly not fortuitously organized random noise or a product of a human predisposition to find faces in noise. The painted wall represents an optically noisy environment, and even though it is not a dynamic energy as found in a video loop, it is chaotic and it is physical, satisfying

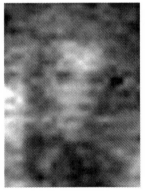

Figure 12-3: ITC Feature like the "Face on the wall"

all of the predictions of the Stochastic Resonance Hypothesis—if you allow the existence of the external force of psi energy.

A discussion concerning cross-correspondence is appropriate here. In cross-correspondence, two or more people or spirit communication techniques will produce information that is of essentially the same content. In this instance, the congregation of the Desert Spiritualist Church conducted a group meditation shortly before vacating the building for the last time. During that meditation, they asked the lady whose face was represented in the wall to follow them to their next church building and to make herself visible to others. That was on a Sunday and at the time, we were unaware of their meeting, meditation and request to the lady on the wall. The following Monday evening, we conducted a Video ITC experiment. Figure 12-3 is a feature that we found in that session. As you can see, the lady from our experiment bears a strong resemblance to the lady on the church wall. When we saw the face, Lisa immediately recognized it as the same lady before finding a photo to compare the two. The timing of this and the resemblance is considered by us to possibly be a case of cross-correspondence.

The Holographic Hypothesis for Video ITC Features

The standard nomenclature used to describe Video ITC features is explained at the beginning of this chapter. One of these, the holographic feature, may point to a very important characteristic of how nonphysical reality is experienced through physical phenomena. This is a relatively arcane concept, so we will explain the elements first.

♦ The "emergent patterns" in individual Video ITC frames tend to contain face-like features.

♦ In actuality, all of the optical texture found in Video ITC frames may contain features, but we are only able to see these features in areas of relative brightness.

♦ Some video frames have hundreds of emergent patterns suitable for transfiguration. However, most of the resulting features are not well formed. For instance, it is common to find numerous faces that consist of little more than ten or fifteen bits of appropriately placed light and dark spots. See Figure 12-4. We sometimes refer to these as "face fractals." The resulting features are

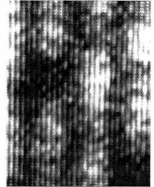

Figure 12-4:
Holographic Feature

sufficiently face-like to make "faces in the clouds" an unlikely explanation, but not sufficiently face-like to prevent some observers from arguing that they are just dots and that the appearance of a face is just an illusion.

There is nothing quite like the experience of loading an almost black video frame into photo editing software, and after a few adjustments in intensity and contrast, finding it to be full of little faces. Seeing so many face-like features, some of which are well developed likenesses of human faces with individual characteristics, drastically exceed our ability to comprehend their significance. Yet, we are certain that we perceive something that challenges our science, pre-telling of a reality that is barely at the fringe of our perception.

♦ These faces will tend to overlap such that they may share the same bit of optical noise for their formation. For instance, a black area used for one face as a left eye may be the right eye for an adjacent face.

♦ Some faces are well defined with sufficient detail for an observer to recognize such details as eyes, nose, mouth, ears, hairline and even clothing. However, even in those that are clearly human faces, the same areas of brightness in the video frame may host several other faces. One face will usually be dominant, but close examination will show that others are also present—some very clear and some little more detailed than a few not so random dots. It is a little strange to see a well formed face, but on closer inspection to find that the forehead of the face contains another, but smaller face.

An example of these holographic features is included here as Figure 12-4, but these phenomenal images do not reproduce well in print and we would like to ask you to visit the AA-EVP website to see more clearly presented examples. The image in Figure 12-4 is a full video frame that has been rotated ninety degrees to the right. Depending on how well this picture reproduces in print, you should be able to see a black field containing numerous lighter areas. When viewed on a computer screen, it is clear that each patch of brightness contains the likeness of a human face. The one near the lower left corner is the best formed. In that patch of light, you can see spots of gray where the chin, nose, cheeks and forehead should be. There are appropriately placed dark spots for the two eyes and a wide spot for the mouth. There is even a suggestion of a spiked head of hair and evidence of a neck. Although not as well defined, each of the other areas of light has similar faces.

Before we attempt to make sense of this holography-like property of Video ITC, perhaps it would be best to describe a similar holographic effect that can be found in the form of physical mediumship known as "transfiguration." Transfiguration features are often fleeting and many changes can pass over the medium's face in a short period of time, suggestive of the same holographic effect we observe in Video ITC. As reinforcement to this comparison, it is interesting to note the common practice in mediumship development circles to sing, or in some other way bring sound into the room that is of differing frequencies. The objective of introducing these sounds is to change or lift the "vibration" of the room to bring everyone together in harmony and make it easier for the communicating entities to cause physical phenomena. People who use this technique have reported to us that they find that the entities seem to prefer some songs to others. This introduction of sound is not unlike the way we fiddle with the video loop to improve image quality.

So that we will be able to make notes for later use, we sometimes record "brainstorming sessions." In one such session, in which we were discussing this relationship between singing to enhance phenomena and the use of background noise in EVP and Video ITC, we recorded the EVP, *"There's a synergy."* Thus, we might assume that the communicating entity felt that we were on the right track.

The implication of these comparisons is that, by understanding one form of the phenomena, we might better understand another form. Is it possible that the holographic-like features we find in Video ITC point toward a principle that describes the nature of nonphysical reality? Are the people who argue that reality is holographic in nature correct? In hypotheses depending on quantum holographics, for instance those offered by Edgar Mitchell,[24] it is held that the psi phenomena we study are actually a natural product of quantum physical principles. To quote Dr. Mitchell, "... recognition that the quantum hologram is a macroscale, non-local, information structure described by the standard formalism of quantum mechanics extends quantum mechanics to all physical objects including DNA molecules, organic cells, organs, brains and bodies. *"*

It is instructive to conduct a search with an Internet search engine using "hologram" as a key word. Doing so reviews hundreds of attempts to compare consciousness, psi phenomena and God to holographic principles. Obviously, many people[27] have observed the holographic-like behavior of some nonphysical phenomena. The primary source of this observation appears to be the "nonlocal" behavior of information in nonphysical reality. By this, it is meant that a person can psychically access the same information from any location. Also, if God is in everything, then God is nonlocal.

Video ITC features are not necessarily "nonlocal" in the sense that you find nonlocality in holographic imaging. What we observe in Video ITC are features that seem holographic-like in nature. Since we are rendering multi-dimensional video noise on a two dimensional surface, it is entirely possible that the holographic-like features are also nonlocal, but that the video display medium does not support direct evidence of that nonlocality. Much more research will need to be conducted in this area, but for now, it is important to note that we feel that we see some evidence in Video ITC features supporting the contention that reality behaves like a hologram.

It is also important to note that we feel that the holographic model for reality does not explain every aspect of reality, as some researchers propose. Holographic principles will probably help us explain how things work, but they do not appear to be the explanation for the existence of Self or the survival of Self beyond physical death.

This point is sufficiently important to state it in a different way. The Quantum Holographic Universe Hypothesis may eventually explain the mechanism or operation of reality; however, no version of this hypothesis, in our opinion, has effectively explained the presence of self-aware intelligence or the essential elements of the Survival Hypothesis. When the models depending on quantum holographic principles are considered *with* those arguing survival, the resulting view does explain the evidence we find in EVP and ITC. Further discussion of this hypothesis is provided in Chapter 11.

Synchronization Artifacts

The last technical issue is the apparent difficulty communicating entities have in lining up all of the optical bits of color, light and pixels to form their likeness. As we have said, the video loopback approach does terrible things to any bit of order that may exist in the data stream. Forming a recognizable likeness of a person or animal must be a true challenge for whatever intelligence may be at work.

Common synchronization artifacts we see are apparent echoes, in which a primary feature appears to be at least partially repeated nearby. One side of a face may be misshapen, as if the screen had suddenly moved down a little while the face was being formed. It is also all too common for faces to be partially obscured by optical noise.

One example on the AA-EVP website is that of a man's face, but on close examination, we see that he has two pairs of cheek bones. This may be because it is two different faces, but we consider this feature to be an example of an echoed image. It is as if the entity began forming an image of the left side of a face, lost synchronization with the video signal and ended up finishing the rest of face slightly shifted to the right resulting in a face with three eyes, three cheeks and two noses.

With careful examination, you will see that most of the examples provided in this book have some amount of distortion or obstruction due to optical noise. The indication is that, just as we are, our spirit team is still learning how to use this media. Our belief is that, in time, our team will learn to improve the quality of the images and we are beginning to see evidence of this.

Similar problems with synchronization are evident in other forms of paranormal phenomena. For instance, people who draw human faces based on their psychic impressions in mediumistic art sometimes complain that the entity they sense, and are depicting in the drawing, will occasionally change in the middle of a drawing. These unexpected changes result in a drawing that depicts two different people within the same face. For instance, the face in the drawing may have a young man's upper face and an old man or woman's chin and mouth. In effect, the artist and the entity lost synchronization. The man with three eyes that we described seems to be a Video ITC equivalent of this problem.

What we Find in Video ITC Features

Most of the features we find in this method of Video ITC are human faces. The holographic images previously mentioned are usually simple faces. Insert features tend to be whole body or from the lower chest up with evidence of a background scene. For instance, we have found a man dressed in the style of sixteenth century nobility with one foot on a low rock wall, which is located in what appears to be a pastoral scene. Unfortunately, the entire frame is so saturated with noise that it is unsuitable for display.

Figure 12-5: Man with ruff

Figure 12-5 offers another example of the period dress we find in Video ITC. In that figure, we see a man from mid-chest up, facing to your right. You should be able to see light reflecting off of his forehead. He may have a beard, and in the color version, it is evident that he is wearing a two-layered ruff that is close under his chin and low around his shoulders like a lapel. Ruffs were popular during the Renaissance.

Beginning with the first man, we began finding people wearing hats. Not just men, we also find women wearing very stylish hats. In fact one of the more interesting characteristics of Video ITC is the period attire of the people depicted in the frames. The "Standing Man" shown at the beginning of this chapter in Figure 12.1 appears to be dressed in either late 1800s or early 1900s attire. He is not a particularly uncommon example of a feature in which you can make out such details as slicked back hair, a mustache, a possible cane, and blazer.

A few of the people we have found look a lot like the "little people" of Irish lore. Actually, we have found a few "elfin" features that are representative of just about every "species" of little people you find in the lore. There are a couple of possible conclusions to be drawn from this. First, we know that Video ITC images are evidence of the existence of nonphysical entities. Therefore, finding elfin-like people in our ITC could argue that they either did or do exist in our physical aspect of reality. Alternatively, they could exist in some alternate aspect of physical reality or in some aspect of nonphysical reality. Could they be devic entities or nonphysical nature spirits that have accepted the appearance we humans have endowed them with over the years?

If the universe holds other life forms, then we should also find evidence of these other life forms in our Video ITC. In fact, we do. An apparent extraterrestrial "person" is shown at aaevp.com. In this feature, we see a person whose body is facing to your left, but who is looking toward you. He has a short forehead and appears to have a bald head. His chin is unusually broad and his head is hanging from its neck in an unusual way. He also appears to be standing in front of a large round window. You must be the judge if this is simply a much distorted "Earthling" or if it is an extraterrestrial.

Yes, we do believe we have found examples that look a little like the gray extraterrestrial as they are often depicted in the media. There are many features we have collected that appear to be of extraterrestrial life forms, but because we do not know what these other life forms look like, we are forced to classify them somewhere in the category of "strange creatures." More interesting though, are the ones that look human, except for some small detail. For instance, as shown in Figure 12-6, we have found a man who appears to have an armored chest plate, some form of ruff around the neck and a very blue face. One of the interesting characteristics of Video ITC is that the

Figure 12-6: Extraterrestrial?

images are often appropriate in color, which is no small thing for such chaotic video circumstances. So, in this figure, the chest plate seems to be the right color for something metal. The color of the ruff is plausible as well. Does that mean that the man's blue face is correctly colored? We cannot know for sure. If it is, then we might be permitted to believe that we have another image of an extraterrestrial.

There are examples of many different animals in our Video ITC. For instance, we have seen cows, horses, dogs and cats, a wolf and many other types of animals. There have even been a fish or two. For those of you who love animals, and who wonder what becomes of them when they transition, we are happy to say that it is rather common for us to find people with their pets. We find many people holding their dog or cat, or closely situated with an animal. It is as if people are proudly presenting their beloved animal. Figure 9-1 is a good example of this. In it, you should be able to see what appears to be a nineteenth century military man that is facing a little toward your left, and holding a dog in his right arm. The dog is at the left of the frame and slightly below the center. This image is also an example of how the features are sometimes distorted. The little Terrier dog's image is appropriately shaped but the man's head is distorted.

From the beginning, we have found people together inside buildings. One frame looks like individuals in a large room in which the seats were tiered like an auditorium. The room even appeared to have windows above the top tier of seats. Another indoor image showed people sitting around a circular table in some sort of gathering. Other scenes show people outside with what appears to be landscapes, just as you would see on earth. In the future we hope to find more of these with suitable definition for printing.

The ITC images that we are finding have improved and we are receiving more communication regarding the processes involved. Perhaps these improvements will continue as this bridge between worlds is strengthened.

Summary of Conclusions about Video ITC

Video ITC is another proof of personal survival and a greater reality. This greater reality is not just populated with those who have lived on earth. It apparently contains animals and all other life forms, as well. If one looks over the many ITC images that have been transmitted to researchers all over the world they will see water, mountains, plants and even towns. The ITC images also show life forms that we do not recognize as ever having been on earth.

We have attempted to show that Video ITC is a complex and robust form of physical phenomena. While EVP offers recognizable voices saying meaningful things that clearly affirm our survival beyond physical death, Video ITC offers a tantalizing glimpse into the community of life that inhabits that greater reality.

Chapter 13

How to Record Video ITC

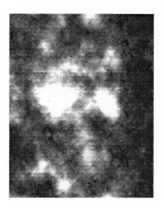

Instrumental Transcommunication (ITC) is the term used to describe spirit communication through the use of technological, generally electronic, devices. In Video ITC, a video camera is used to record paranormal images. The presence of these images cannot be explained with currently known physical principles. Other than EVP, the use of a video camera and a television set or computer monitor configured in a video loop, is probably the most repeatable method for studying these phenomena. Video ITC does require more hardware than does EVP, but the results can be very dramatic.

For the record, the Video ITC technique described here is known as the Klaus Schreiber[19] method for Video ITC. The technique of connecting the output of a video camera to the input of a television set, and then taping the video noise that can be seen on the television screen, was apparently first developed for Schreiber by Martin Wenzel. It was a modification of this method that led to our first success. His suggestion was to place the camera just a few inches away from the television screen and then focus beyond its surface.

Although a motion camera is used in Video ITC, the final results are "still" images. As with EVP, the Video ITC experimenter generally does not see the phenomenal features while they are being recorded. The experimenter expresses the desire to receive these images and the communicating entities comply by impressing features in the video noise. It is possible to ask a particular entity to place his or her image in the Video ITC. As previously described, we have succeeded in requesting Tom's father to appear in the middle of a video frame.

Using the Video ITC Technique

The following technique is designed to enable anyone to replicate the experiments we have been conducting in Video ITC. This is not the only way to conduct Video ITC experiments. It is a technique that has worked for us and has worked for other beginning Video ITC experimenters.

The essence of Video ITC is the creation of background energy in the form of optical noise via the feedback circuit between display and imaging devices, and then recording that noise in an effort to obtain "snapshots" that can be individually examined. The snapshots are the individual video frames. An alternative source of optical noise will be discussed at the end of this chapter.

Equipment: There are numerous ways to conduct Video ITC experiments. For instance, the beginning setup we describe here uses only a portion of a television screen, but at present, we are using nearly the entire screen for most of our experiments. You may record with the video camcorder or use something like a security camera or web camera that has no recording capability. You may also use a stand-alone video tape recorder and/or a computer for the recording device. You should be able to use a computer monitor rather than a television set. Researcher Pascal Jouini did this and recorded faces. Some people prefer to record in black and white rather than in color, and some experimenters will remove the tuner stage of the television set to assure that no broadcast channels are involved. These are options which you may wish to explore as you gain experience and depending on your available resources.

Please note that we are not experienced with digital camera recording and that these instructions are specifically applicable to analog cameras. We also have no experience with the use of a digital monitor. A digital video camera and/or monitor may work, but since we do not know exactly how or where the features are injected into the circuit, we cannot say how the use of digital technology will affect the results. We do know that even the slightest change in camera settings and position results in a dramatic change in the resulting features. From this, we believe that two cameras that are of the same manufacturer and model, and that are set up with identical supporting hardware, will produce different results.

Just as we suggested in the chapter on recording EVP, if you want to experiment with video ITC, try to do so with the equipment that you have available. The required equipment can be expensive for a person to purchase it all at once.

The following list of suggested equipment and software will be explained in detail, and then the procedure for using the items will be described so that the steps can be clearly viewed in the proper order.

Video Camera: Any analog camcorder should work just fine. We use a Canon[20] 8mm ES2000 analog camcorder. Because the camera is analog, we require an input capability for our computer that includes S-Video or Composite Video. For us, this input capability is currently a Pinnacle Systems[21] Studio DC10 Plus analog video capture card. The camera we are using has a major flaw for Video ITC in that we cannot lock the focus and zoom where they were last set, because these features reset to automatic mode when the power is removed. When we have a particularly exceptional session, we would like to conduct another session with exactly the same settings. The Cannon video camera makes this impossible. So, if possible, you will want to use a camera in which you can return to the same focus and zoom points each time you turn on the camera.

It should be noted that the video interface is important. The objective is to record as many pixels as the equipment will allow. The number of pixels in the frame in a television set, and what most camcorders will record, are fixed at the industrial standard for broadcast video display. However, depending on the interface you are able to use between the recorder and the television set, and later, between the recorder and the computer, there can be a difference in the overall resolution (pixels/inch) of the final video recording. The Composite Video interface, which is usually the yellow input jack near two audio jacks, will not support as high a bandwidth as will S-Video. The USB port on a computer will support only about half the resolution as will the S-Video port. The newer, USB 2 and Ethernet ports should offer resolutions comparable to an S-Video port, if you are using a digital video camera.

One exception to the need to use the highest bandwidth interface available is the interface between the camera and the display for recording. We are unable to achieve the desired feedback pattern when we connect the camera to the Video-In of the television set using the S-Video port of the two devices. Consequently, we use the Composite Video ports during recording and the S-Video ports between the camera and the computer during the capture of the frames.

Most digital monitors are configured to operate at a much higher resolution than televisions or video cameras. If you can take advantage of this by using a video monitor in place of a television set, the higher resolution may be beneficial. We know of people who have used computer monitors for Video ITC, but so far, we have not received a definitive report that the increased resolution helps. In fact, we know that Pascal Jouini experimented with a computer monitor to prove that the images were not bleed-over from television signals, got positive results, and then returned to the use of a television screen.

The small webcam that often comes with computers is a low-resolution device. We have attempted Video ITC with one of these without success, but we should also note that we did so using a liquid crystal computer monitor. These two digital based devices combined with the low resolution of the camera may have been the problem. In principle, a webcam should work. Perhaps we simply did not find the correct combination of hardware and settings.

As we have stated, we have not used a digital camcorder. Again, in principle, a digital camera should work, but we want you to be aware of the possibility that the entities cannot work with digital formats as readily as they are able to work with analog. Based on what we know today, we believe there needs to be at least one analog component in the video path. However, we simply have not had the time or resources to explore this.

When conducting video ITC experiments, we use our camera in the color mode of operation. However, some experimenters say that it is easier to see the features in black and white, and so, they convert their color video into black and white after the frames are captured.

Video Tape: Any video tape that works with your camera should be fine. Always use a fresh video tape. You will be marking the beginning of each session with a one or two second exposure on a focus page that includes details of the experiment. Since each session is usually less than a minute long, this should allow the use of a single tape for many sessions, assuming that a separator, such as a focus page, is used.

The used video tape is saved as an archive, but we depend on saving the captured video onto compact disc for our primary archive. It is especially important to save the video tape record of experiments if you conduct experiments to prove the reality of these phenomena to others.

Environmental Modifiers: Some Video ITC experimenters who use the video loop method also shine various colored lights, such as ultraviolet or fluorescent lights, at the television screen during experiments. The idea is to bias the feedback loop toward one optical frequency or another in an effort to provide better energy for feature formation. This provides a form of background energy which is much like adding background sound for EVP, making the similarities between EVP and Video ITC even more evident.

Although not shown in Figure 13-1, we usually use a sixty watt incandescent lamp which is on the floor below the television set and facing the floor. This provides a small amount of optical energy as a bias to the video loop. The room is otherwise darkened, so the light also helps us see to maneuver around the camera and tripod.

Television Set: The television that we currently use for experiments is a twenty-inch color Sony, but we have also had success on a thirty-four-inch Panasonic set. The size and make of the television does not seem to be important. The distance of the camera from the television and the size of the set will determine how many television screen pixels are in the exposure window.

As we understand this technique, the objective is to have as many pixels in the exposure as possible. However, considering what happens to the pixels as they pass through the video loop, the number of television pixels the camera detects may not be important. Remember that there is usually the same number of pixels in both the camera and on the television screen. When the camera is placed very close to the television screen, only a small percentage of the pixels on the television screen are recorded, but they are always recorded with the same number of pixels in the camera. The result is that the pixels of the television screen are, in effect, enlarged each time they pass through the video camera and back to the screen to be recorded again. It would require just a few passes through the video loop to enlarge the screen pixels beyond recognition. More or less television pixels should not be a critical factor except for the potential sharpness of ITC features.

Because the focus of the camera is beyond or in front of the screen, even further obliterating the screen pixels, the size of the set should have little to do with the results. We have found that different television sets produce slightly different results, but we believe that this is largely due to the electrical circuit causing a shorter or longer feedback loop. Perhaps it is best to think of the television set as nothing more than a loopback device that functions as part of an optical noise generator.

By the way, should you think that we might be video taping a television program then consider what we have just said in the previous few paragraphs. When we conduct an experiment, any possible television broadcast image on the screen would be quickly obliterated in the video loop.

A television set that has a Composite Video-In jack is required. If the television set had a broken tuner, that would be all the better. If you are handy with a soldering iron, wire cutters and a screwdriver, it is possible to disable the tuner section of the set. However, we would advise that you do this only if you are afraid that you might pick up a broadcast signal in the video loop. Also, please do not attempt to make changes to the electrical circuit of a television set, or any electronic component, without proper training in working with electricity.

Tripod: The camera needs to be positioned so that a portion of the television screen will be exposed. Also, you will want to be able to change the camera distance from the screen and be able to pivot the camera away from the screen to set the focus. It is helpful to be able to return the camera to the same location for repeated experiments. Anything that will hold the camera in place will do; however, a camera tripod will facilitate these positioning tasks.

Computer: To analyze the video we use a Windows based Computer. Any well-equipped computer should do. Look for plenty of video and hard disc memory and a spare circuit card slot if you are using an analog camera. A read-write CD drive for storing all of the ITC images you will collect is extremely helpful. However, the final video ITC files are usually small enough to be saved onto a floppy disc. If you are using a digital video camera, you will want to make sure that you have the appropriate digital input port, depending on your camera. Caution, the early USB ports will not support the full resolution of the video camera. The newer USB 2 ports will do so in most computers.

Video Capture Card: To capture the video into the computer we use the capture card that came with Studio AV by Pinnacle Systems.[21] It came with analog inputs and video editing software. Once the video has been transferred, or "captured," into the computer, it is more or less "standardized," therefore any method of video capture should work. If a digital video camera is used, you will probably be able to use one of the computer's existing input methods. As noted above, avoid using the early USB ports because of their low bandwidth.

Video Editing Software: Pinnacle Studio Version 8[21] is our current video editor. However, there are several very good video editors on the market. Whatever you use, it should have the capability of displaying one frame at a time and "grabbing" selected frames as still photographs. The ITC images are found in the grabbed frames.

All of the consumer video capture and editing programs that we have reviewed have a relatively small viewing window for displaying individual frames. Video capture cards that have a Composite or S-Video output port will allow what is displayed in the review window of the editing software to be simultaneously displayed in a television set. It is possible to make do with the small review window, but it may be worth the effort if you can arrange for a larger screen, as it would help you eliminate marginal frames. In the Studio DC10 software, there is an output selection option that allows you to select between Composite and S-Video inputs and outputs. Make sure that these are correctly set for video capture. If an external monitor is used, make sure that the output settings are correct.

There is one final note about using the video editor software. It is usually possible to adjust various parameters, such as intensity, color balance and contrast in the video frames within the video editor software. Experience indicates that it is best to reserve these editing processes for the photo editing software, because, often, very small changes can result in dramatic changes in what can be seen as ITC features. Features can be seen better in the photo editor, whereas making changes to the frame in the video editor may cause you to miss important but subtle features.

Photograph Editing Software: You will need the capability to edit photographs. Adobe Photoshop Elements 2.0[22] is the current software we are using. In the past we have used Adobe PhotoDeluxe Business Edition with good success. Some of the software that comes with digital cameras provides very good image editing capabilities, and there are many photo editing packages on the market that should work just fine. Beside the ability to crop, rotate and size images, the main thing to look for is the ability to change the contrast and brightness of the image. Some of the ITC we have collected began as a grabbed frame that was almost completely black. The photo editing software made it possible for us to lighten and improve contrast until we had a relatively clear image.

To preserve detail, work with as high a resolution (number of pixels per inch) as you reasonably can when editing the frames. The resulting images can be saved as low resolution JPEG file. When working with our images, we usually use two to four hundred pixels per inch. When emailing a file or posting it to the web, we save the features as a JPEG file at seventy-two pixels per inch.

Color, and Black and White Printers: For printing we use a LaserJet for black and white printing and a color Inkjet for color printing. Using these printers, we have found that many of the relatively good ITC features simply do not print well. They are typically very dark against a dark background and the printers do not have a sufficiently high resolution to reproduce them. Thus, there may be some difficulty sharing your features via printouts. For public presentation, we use either a slide show capability on our laptop computer screen or a digital video projector driven by our laptop.

Sound Editing Software (Optional): You may wish to examine the sound track, made during the video recording sessions, for EVP. If so, sound editing software such as Audition[18] should be available. See Chapters 10 and 11 for a detailed discussion about recording EVP.

Equipment Setup

As we have previously stated, these instructions are a guideline that can be modified as you gain experience and depending on the type of hardware that is available to you.

Camera-Television Connection: The television should be positioned so that the video camera is a short distance in front of the screen, preferably on a tripod. The output of the video camera should be connected to the input of the television set via the Composite Video-In jack, or other connection, such as S-Video, depending on your capability. The television should be set to receive input from that jack. With the television and camera turned on, you should be able to see the output of the video camera on the television screen. Thus, there is a video loop in which the camera "sees" its own output. To double check that there is a video loop, zoom the lens out so that the television cabinet can be seen in the camera viewfinder. Repeated copies of the television set should be seen on the screen .

It is possible to insert a video tape recorder into this circuit if a camera is used that does not have recording capability. The video tape recorder can also be used to view the images on the television set, frame by frame, if you do not have the equipment to download and view them in a computer. If an ITC image is found, you could then take a picture of it with a digital or film camera. We have included a video tape recorder in the loop as a delay device and found little change in quality or quantity of features.

Camera Position: The camera must be pointing at the television screen so that there is a video loop, but there are no hard and fast rules as to how close the camera should be. During various experiments, we have tried the camera within three inches of a twenty-inch screen and as far back as three feet. Since we set the zoom at about forty percent of maximum in, we are "looking" at from four or five inches of the screen up close and almost the whole screen when further back. The features we have collected have been generally more detailed when collected with the camera close to the screen but they tend to be smaller and more pixilated. The features tend to be larger when collected from a greater distance, but they also tend to be less well defined. The latter are preferable because the pixels are less intrusive.

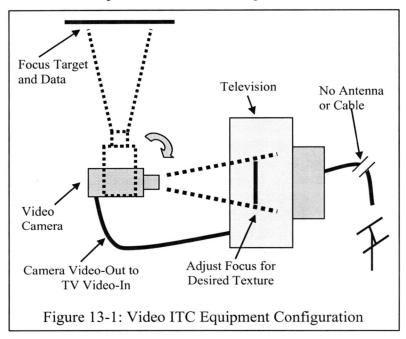

Figure 13-1: Video ITC Equipment Configuration

We have also tried positioning the camera at different angles in relationship to the screen. If a mental line is drawn straight out from the screen and perpendicular to the plane of the screen, then we have positioned the camera at about ten, twenty and forty degrees with good results. Some experimenters also tilt their camera to ninety degrees so that they see the screen sideways in the viewfinder and have had good results. This approach does produce an interesting feedback pattern but the results were not as good for us when we tried this. However, our poor results tilting the camera might simply be due to our team on the other side not having sufficient time to adjust to the different configuration. Since our results seem to be steadily improving at three feet back with a normally mounted camera, we went back to that.

To begin your experiments, we suggest that the camera is placed about three inches from the screen. We feel that this will give the highest probability of collecting features within a short period of time. After conducting successful experiments and finding the phenomenal features, it is reasonable to experiment with different locations for the camera. As just mentioned, our preference today is working with the camera set three feet from the screen.

Camera Focus: The camera should be adjusted to about the same focus each time the equipment is set up so that you can repeat experiments. One way to assure that the camera's focus is at about the same setting each time, is to place a piece of paper with the session information written on it, at the side of the television set. With this arrangement, it is possible to pivot the camera to focus on the target paper.

Begin by placing the camera where you will first record, turn off the auto focus and then adjust the camera's focus and zoom until there is a cloudy texture flashing on the television screen. This is done by focusing past the screen. Sharply defined specks of light, moving from the center of the screen to the edge, will probably be visible at first. As the defocus of the camera is moved past the surface of the screen, these spots of light will dissolve into a moving field of bright cloudy texture interrupted by dark flashes. As we have found, this is in the "range" of the optimum video texture for the formation of the phenomenal images. Once you are satisfied that the camera is properly focused, pivot the camera away from the screen and position the target paper so that it can be seen to be in focus through the viewfinder. Find a way to mark the location of the paper and camera so that you can return them to the same location for future experiments if the current experiment produces good results.

The camera we use to do our experiments might behave differently than the one that you will use. When discussing with another researcher that we saw the cloudy texture in a bright white followed by dark flashes he remarked that he saw this in blue. The cloudy texture that we see is a combination of grays, browns, reds, blues and white.

General Camera Settings: How the options within the camera are selected is not critical; however, note the settings so that experiments can be repeated. The "Portrait" exposure preset is what we prefer to use. Each preset will tend to produce a slightly different chaotic image on the television screen, so experiment to find the one that gives the greatest amount of "cloudiness" between light and dark extremes. Any possible camera control information that is displayed on the television screen, such as "Record" and "Auto Focus," should be removed, either with a setting in the camera or with the remote control that may have come with the camera.

Procedure

Once again, we need to state that the following procedure is a beginning point, which can be modified as you develop your personal technique. Steps 1 and 2 are designed to assure that you can control the focus of the camera. If the focus control of the camera can be locked, you will not need to use these steps, other than to record a second or two of a target paper to mark the beginning of the next session.

1. **Preparing the Video Loop**: For the first experiment, turn on the equipment and position the camera about three to five inches in front of the television screen. Connect the Video-Out of the Camera to Video-In of the television and select Video-In on the television set. Aim the camera at the television set and slowly adjust the focus, and zoom until the dark to light flashing is visible on the television screen, with cloudy or foggy texture in various colors during the bright flashes. The focus should be six to twelve inches beyond or past the surface of the screen when the equipment is ready to record. The most important objective is to see swirling clouds.

   We are emphasizing the focus because we feel that focus and zoom are the most critical determinants in creating the necessary optical energy for feature formation. In our opinion, it is entirely possible to conduct a successful experiment, and to follow it with a less successful experiment because the focus and zoom might not have been returned to the same settings. A second feature of being able to accurately return to settings from a previous experiment is the possibility of being able to gradually change the focus over many experiments to optimize the settings.

2. **Preparing the Focus Target**: As previously described, pivot the camera to point at a target paper that has identifying comments

for the current experiment. Position the target to be in focus as seen in the camera viewfinder, and then mark that location. Record the target paper for about one second. If the experiment is successful you will be able to focus on your target paper in your next session so that you can return to the same setting.

Turn the camcorder back toward the television screen.

3. **Preparing Yourself:** Follow the same preparatory process you use for EVP sessions. Remember that there are nonphysical entities who are present and able to "witness" your activity. Consider using the same music each time as a "signature" or "signpost," indicating that preparation for a session is underway. Also consider conducting a short meditation or prayer.

After meditation, we change from music to the background sound that we use for EVP. The background sound of white noise is used because we always listen to the video sound track. Also, an IC recorder is usually recording during the sessions. This, of course, is optional. Speaking out loud, we talk to our team about the last session and discuss the various successes or failures of that experiment. Asking for their assistance in bringing the images through, we announce that we are going to begin the experiment. This is all done just as if they were standing in the room with us. Each session brings different discussions and questions. Ask for information on how to improve the experiments, and for specific people to show themselves in the video.

If working with another person or a group, discussing and agreeing on what sort of features you would like to receive before the experiment may be helpful. After deciding this, consider conducting a joint meditation with each person focusing on what the group wishes to receive.

The experimenter is part of the circuit and we feel that meditation and/or prayer helps bring the experimenter into a more balanced state. This helps to focus the experimenter's intention and better helps those on the other side create a link to the experimenter.

4. **Conducting the Experiment:** Speaking out loud, tell the entities that you are about to begin recording. State what you wish to see in your video frames, and perhaps, offer feedback about the pre-

vious experiment. Turn on the equipment and wait a few seconds for the feedback loop to stabilize. Record for about thirty seconds. You may record longer, but remember that the camera will record around twenty-nine frames a second, and that thirty seconds represents a large number of frames. Turn off the video camera and verbally thank the entities for their help.

Evaluating the Video

Transfer the resulting video into the computer and begin the process of looking for ITC features. This is a frame by frame process. As you can see in the screen print in Figure 13-2, the video editor we use provides a small viewing window which is at the right of the picture. With close examination of the video frame that is displayed in the viewing window, you may make out a pattern of light and dark that looks a lot like a distorted human face. It is in the middle of the frame.

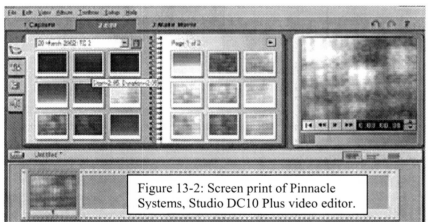

Figure 13-2: Screen print of Pinnacle Systems, Studio DC10 Plus video editor.

The objective is to look for irregularities in each video frame. There are a number of examples at aaevp.com that should help you begin recognizing frames that may contain features. When going through the frames one by one, we "grab" just about all of the frames that seem to have unexpected deviations in texture and then examine them more carefully using the photo editor. The result can be hours spent examining frames with the photo editor.

Analysis of Video Frames for ITC Features

There is a discussion of how we humans see ITC features, and their validity as phenomena, in the previous chapter. The question is important because there are characteristics of our mind that do influence how we see things and that could lead us to mistake a natural artifact of the Video ITC recording process as nonphysical phenomena. This capability of mistaking random optical noise for familiar objects is sometimes described as the "Faces in the Clouds" or "Rorschach Test" explanation. The bottom line of our comments in the previous chapter was that we feel extremely confident that the images you will find are a valid form of nonphysical phenomena.

Here, we will concentrate on the steps necessary to recognize and find ITC features in "grabbed" video frames. We will base our comments on the Adobe Photoshop Elements 2.0[22] software that we use; however, any good photo editing software should suffice. Besides the usual copy, paste, new and save, the primary capabilities needed in a photo editor include:

♦ **Size:** The ability to set the width and height of the image, and the number of pixels per inch.

♦ **Crop:** The ability to select an area of a larger picture and either copy it into a new frame or delete the portion of the original frame that is not wanted.

♦ **Save as JPEG:** This format works well because it is easily sent over the Internet and takes up little disc space. The negative is that the format loses some resolution that cannot be regained, so consider keeping your original "grabbed" frames on a compact disc for future reference.

♦ **Color to Black and White:** Some features simply do not show up well in color. We prefer to keep the color whenever possible, but there are times when it becomes a choice between having color and being able to see an image. Using the editing software to remove all color can sometimes make features more clear.

♦ **Contrast and Brightness Control:** Some frames are nearly black when analysis is begun. By changing the contrast and brightness together, we can usually bring out features that would otherwise be invisible.

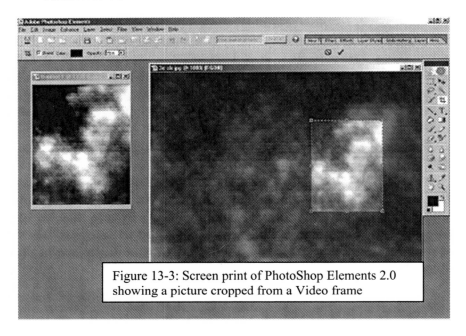

Figure 13-3: Screen print of PhotoShop Elements 2.0 showing a picture cropped from a Video frame

Figure 13-3, which is an Adobe Photoshop Elements 2.0 screen print, gives an idea of the process used to find ITC features in video frames. The image at the right is a frame that displayed texture, and therefore was "grabbed" during our review of video. An area to the right showed a possible image. The selected area can be seen as a lighter rectangle that was copied into a new canvas to create a second file so that we could see if the feature was worthy of being saved. As you can see, the image at the left in Figure 13-3 looks a lot like a child's or woman's face. Brightness and contrast of the feature was then adjusted for clarity.

After looking at the frame in the normal position, we then rotate the frame in ninety-degree increments to check for images at other angles. In this particular image there were the beginnings of several faces at other angles but none was worth working with.

When studying these frames in the photo editor, make a point to stand back from the computer screen from time to time. There have been numerous instances in which we were looking at a frame that seemed to have nothing on it, but when we left the computer to do something else, we saw a major feature on the computer monitor from across the room. The features are nearly obscured by video noise, and it is often the case that a change in our perspective has revealed features that we might otherwise have missed.

Attention may be focused on one image, but often when we simply refocus our eyes, other features will be discovered in the same location. This is the case with the holographic-like images. It is as if there is so little available optical energy suitable for impressing their likeness that the communicating entities are "piling on" wherever they can. Again, remind yourself to refocus your attention from time to time as you look for features.

Photographic ITC

Since the features form in the optical noise, it is reasonable to expect that other forms of optical noise will also produce phenomena. In fact, researchers have found that ordinary photographs have images hidden away in areas of optical noise. Features can be found in the medium intensity areas of photographs, such as a wall or ceiling in the background of a flash photograph.

Jutta Liebmann, of the German VTF, photographs her television set while programs are being shown. She waits for scenes that have a lot of optical texture, such as clouds or an aerial picture of water. She has found some very interesting ITC features in this way.

Things to Consider

In practical application, there is a difference between Video and Photographic ITC and spirit photography. Spirit photography is generally a photograph of an apparition or energy form. The entity is said to have been photographically "captured" in an otherwise mundane scene. In ITC, the image is thought to be deliberately caused by the entity through some form of transfiguration of available physical energy. Even in the instances in which the face of an entity is in a photograph or video frame taken of a mundane scene, it is understood that the entity has somehow caused the image to be injected into the recording medium.

The ITC methods we have described in this chapter are based on the creation of optical noise with a video feedback loop. In effect, you are taking photographs of the resulting noise. And indeed, it is possible to simply create a video loop with a cheap camera and television set, and take photographs of the television screen with a still camera, as in the Photographic ITC method.

"Grabbed" video frames that are BMP files are around nine hundred kilobits in size, but when the same file is saved as a JPEG file, it can be as small as thirty kilobits. However, there is a lot of difference in resolution between the two formats. It is best to search the BMP file for images, edit those at a relatively high resolution—say two hundred pixels per inch—and then save the result as a JPEG file. In our system, JPEG files are stored in our computer and the raw video file and BMP files are stored on "write only" compact discs. As a rule, we save features as seventy-two pixels per inch JPEG files if we are going to email one to a friend. We use a higher resolution if we think it is necessary to preserve detail.

Finally, remember that we are all learning how to experiment in this area of ITC. What we know about Video ITC has been shared in this book. In turn, we would appreciate hearing from you about how your experiments go and if you find a better way to bring the images in. It will be through such sharing of information that we will one day find a way to change Video ITC from an experimental art to a reliable tool so that others may benefit.

Chapter 14

Closing Thoughts

Reality is said to be continuous, meaning that its entirety is governed by a common set of principles and that change is gradual from one aspect to another. From this, it is reasonable to believe that we can know something of the nature of the worlds on the other side of the veil by looking at our world. It is widely accepted that who we are here is not very different from who we will be after we transition to the other side. Contacts through EVP and ITC have shown this to be true.

The first paranormal "voices," giving hints about what life was like on the other side of the veil, had to come through the air as direct voices. These voices were heard in the presence of shamans and physical mediums. Electronic voice phenomena only came about when the technology was available and history shows that EVP did not really gain any attention until the 1940s. Even today, few people know about these voices from other worlds.

The ITC images have even less of a history. Perhaps the first images could be considered to be the materialized forms of people and animals seen and studied by researchers during séances. Humankind became aware of these paranormal images as our technology advanced to include the ability to take pictures. The first ITC images began appearing in the 1980s with the use of video and televisions, and then later through computers.

In the 1990s a group of scientists and researchers from the other side teamed up with a group of sitters here on earth, who lived in Scole, England, to develop a new kind of energy to produce phenomena.

Looking at these contacts it is obvious that our friends on the other side are as interested in communicating with us as we are with them. The paranormal contacts through devices show progression and advancement on both sides. Physical death does not obscure love or the desire to help loved ones who are still in the flesh. Thus, we know that everyone who conducts research and experiments with these phenomena have counterparts on the other side who are working just as hard.

If you have the time and interest, consider following some of the simple directions set forth in this book to see if you can contact the other side. It is possible, and it will be one of the most fulfilling things you can do in life. At the very least, keep in mind the meaning of some of the messages from our nonphysical friends that we have described here.

You do survive and you will, sooner or later, find yourself on the other side. Be prepared. Learn what you can now so that you will know what to expect when you do make that most interesting transition. Greet the life change as a great adventure. You can, you know.

When you do experiment with EVP and ITC, believe in the bridge model that we have described and know that the veil may be near to fading away. For us, when we go to the other side, our work will continue as it has in this lifetime. We will continue working to build that bridge because we feel that it is the most important work we can do.

It is exciting to think about what the future will bring. This is a relatively new phenomenon and has only come about through our world's advances in technology. At the same time we can see that the other side has made advances in communication and has even used new kinds of energy to do this. One thing that is certain is that the communication between worlds that has already taken place proves there is no death and there are no dead.

Finally, we wish to thank you for taking the time to read this book. Your purchase of the book will help us continue the work of the AA-EVP, and for that, we also thank you.

Environmental Influence on EVP

We believe EVP and ITC are made possible, at least partially, because of the mediumistic ability of the experimenter. Everyone is inherently a medium. Some people are just naturally more able mediums than are others in much the same way that everyone has some athletic ability but some people are better athletes than others. Just as athletes can train to improve their ability, so too can individuals train to improve their mediumship ability. Following this analogy further, the day-to-day ability of an athlete changes somewhat, depending on such factors as the weather, the athlete's eating habits and how rested he or she is. An individual's day-to-day mediumistic ability is also influenced by such factors as the person's ability to focus on the task at hand, attitude about the situation and his or her energetic agreement with the situation. There are actually many environmental influences in mediumship, but we would like to point out a newly identified influence that may eventually lead to major breakthroughs in our understanding of psi phenomena in general.

To quote a report written by Dr. S. James Spottiswoode,[39] "Evidence has been given to support a relationship between the local sidereal time at which an anomalous cognition experiment occurs and the resulting effect size. The primary association is an approximately fourfold enhancement in AC effect size at 13.5 h LST. *(Anomalous Cognition (AC)" seems to be a new term for Extra Sensory Perception (ESP)—Editor)* This association was found in one large data set and confirmed in another, each set comprising AC experiments with a range of free response protocols, from different laboratories and investigators. It is likely that the increase of effect size for AC trials occurring at 13.5 h LST is real, replicable across different laboratories and occurs in the diverse protocols of the ganzfeld and remote viewing experiments."

Dr. S. James P. Spottiswoode has published a number of studies reporting his growing understanding of the way environmental conditions, such as geomagnetic weather and solar flare activity influences psi functioning. In general, it is clear that solar flare activity has a direct influence on geomagnetic activity, and that in turn, has an influence on psi function. The evidence is still not fully understood, but it appears that psi functioning is slightly improved when the geomagnetic activity is greater.

However, when it comes to sidereal time, the story is very different. Dr. Spottiswoode has examined the rather large collection of psychic ability experiments he has conducted over the years to see if they pointed to a relationship between sidereal time and the psychic proficiency of his subjects. There was a direct relationship! He then asked colleagues to conduct a new set of experiments to confirm his conclusions. The results, and therefore the phenomena, were verified.

Sidereal time is star time and a sidereal day is approximately three minutes, fifty-six seconds shorter than a solar day. Thus, Local Sidereal Time (LST) moves backward in solar time about four minutes a day, two hours a month and one day a year. At any place on the planet, at the same LST, the same stars will be seen overhead.

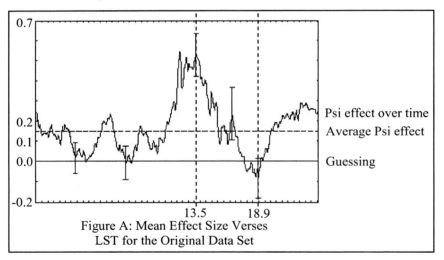

Figure A: Mean Effect Size Verses
LST for the Original Data Set

The essence of the article referenced is that scientists have found a direct correlation between the sidereal time of day and success in psychic ability experiments. The graph shown in Figure A is from Dr. Spottiswoode's article. The graph depicts "Effect Size" on the vertical axis and "Local Sidereal Time" on the horizontal axis. "Effect Size" is the amount of deviation more or less than the expected normal for chance. The horizontal line between 0.1 and 0.2 represents the average of the graph curve. The line at 0.0 represents what would be expected with guessing. The graph is from Dr. Spottiswoode's work, but we have added vertical, dotted lines at 13.5 hours and near 19.0 hours

Psychic ability is real and the problem proving it in the past may be that researchers have been conducting experiments at different sidereal times of day. Almost a six-fold difference in performance of a psychic between 13.5 h and 18.9 h LST is substantial. And remember, 13.5 hours LST changes in solar time each day.

The evidence suggests that there is something near or beyond the edge of the Solar System that is influencing our psychic ability. Of course, experiments will need to be conducted to see if this influence affects EVP collection in the same way as psychic ability. But, remember that mediumship is a form of psi functioning.

The discovery of this relationship suggests an external influence on psychic ability and may explain why psychic phenomena are so difficult to prove. It should be just a matter of time before someone figures out what that influence is and what it is influencing in the human brain. Once that has been accomplished, it should be possible to enhance psychic ability with technology.

In the meantime, you may wish to consider scheduling your EVP and ITC experiments in accordance with Sidereal Time. We have added a sidereal time calculator on aaevp.com and you can download a good clock for your computer at:

www.radiosky.com/sidclockdownload.html.

Techniques for Editing Sound Files

Some of the more important advancements in the field of EVP and ITC have been the availability of affordable and powerful computer software tools which enable anyone with a computer, especially a Windows computer, to review and edit sound and image files. For instance, the increasing popularity of the IC recorders would not have been possible were it not for the ability to load sound files into a computer to analyze and store the recordings. Also, the Internet has enabled experimenters to easily share sound and image files so that others around the world are able to see and hear examples of these phenomena.

These enabling tools can present a formidable challenge for anyone who is not well versed in the use of a computer. One of the more important features of the AA-EVP discussion board, we call the "Idea Exchange," is the possibility that the more experienced experimenters may answer questions for the more computer challenged members. For instance, we have seen numerous messages offering helpful hints concerning the use of sound editing software for recovering an EVP utterance that is otherwise lost in the noise of a sound file.

The use of computer programs for EVP and ITC is more of an art than a science, as each offer a number of ways to solve particular problems. The following paragraphs describe some of these techniques for the more commonly used programs. However, we advise that the experimenter should select a program and simply experiment to see what each feature offers for improving experimental results. Please note that http://aaevp.com/techniques.html includes a growing list of techniques for both EVP and Video ITC editing.

Using a Noise Reduction Process on Sound Files

We have found that if there is a relatively consistent background noise in an EVP recording, perhaps from a noise generator or a fan, Audition[18] can be used to effectively remove that noise. (Note that there are similar capabilities in other programs, such as Acoustica and Gold Wave, which are commonly used by AA-EVP members.)

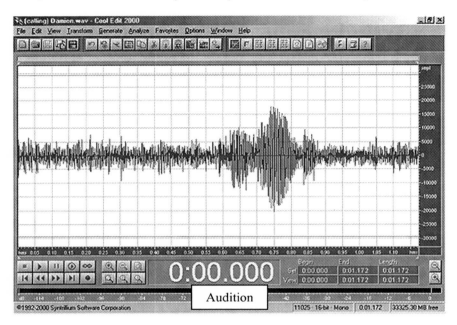

To Begin: Open Audition and start a fresh recording session: Under menu item, FILE, click on NEW and set Sample Rate = 11025, Channels = Mono, Resolution = 8 bit, then click on OK. It may be necessary to experiment with different sample rates to successfully load a sound track into the computer.

Play the sound track into the computer while Audition is in RECORD mode. Try different outputs from the tape recorder. The best is LINE-OUT of the recorder to LINE-IN of the computer. Also, check the SOUND and MULTIMEDIA application to be sure that the correct input jack is active. Look under the AUDIO tab in the SOUND RECORD window and then click on VOLUME. Place a check at the bottom of the volume control that represents the input jack being used.

If the headphone jack is used in the recorder to transfer files, a "Dubbing" or "Attenuating" patch cord may be needed to help match the resistive difference between HEADPHONE-OUT and LINE-IN. These requirements tend to vary from computer to computer.

For Noise Reduction: Once the sound track is visible in the sound editor application, select a few seconds of sound track that has typical noise but no voice (or no suspected voice). Go to menu item, TRANSFORM, and select NOISE REDUCTION and NOISE REDUCTION again. Click on GET PROFILE FROM SELECTION. (If there is not a large enough data sample selected, this option will not be available.) Next, notice that a graph will be displayed and SAVE PROFILE will become available. Save the profile and use it as a "standard" filter for similar background noise. Please note that the saved Profile is specific to the sample rate of your sound file.

When the application has finished, click on CLOSE (do not click on OK) and then select the entire sound file or that portion you wish to analyze. Then open the noise reduction window again and click on OK. This will filter the selected sound track based on the profile of the previously selected noise. (The UNDO may be selected if a SAVE function was not performed. Be sure that ENABLE UNDUE is checked under the menu item, EDIT.)

Since EVP is formed from the noise, reducing the noise may reduce

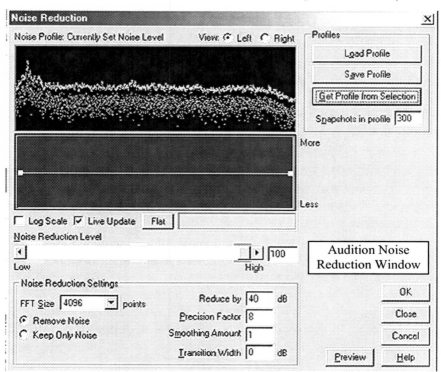

the voice as well. Try different NOISE REDUCTION LEVEL settings. This is a slide selector just under the noise profile graph. This process may need to be repeated to find the best combination.

Depending on the uniformity of the noise on the track, the voices should stand out more from the noise. Please note that this process sometimes induces a "ringing" sound into the sound track. This noise is considered an artifact.

If unable to select a large enough sample of noise without also selecting possible voices, select what you can and open a new sound file. Make sure that the new file has the same settings as the original file. Then repeatedly paste the sample into the new sound file until there is enough to create a profile. Once a profile is created, bring up the file being analyzed, select the sound file and then go to the NOISE REDUCTION window and click on OK.

Audition has many capabilities that will assist in listening to the voices as well as bring them in more clearly. Another piece of software, Clear Voice Denoiser,[28] is only a noise reduction program, but several AA-EVP members have had good success with it. At the time of the writing of this book a free demo version was available on line at www.speechpro.com. The free version has a size limitation and only short segments of a recording can be loaded. Most EVP samples are of short duration; so, if only the EVP sample that needs to be cleaned is loaded the size limitation should not be a problem.

Reversing a Sound Track. Some EVP messages are found on the reverse of recordings. In Audition, to reverse the sound track, select the entire sound track and use menu item TRANSFORM and click on REVERSE. Playing the sound track in this way will be much the same as playing a cassette tape in the reverse. You will hear your voice spoken backwards and then you will hopefully hear EVP messages spoken forwards in the areas where you are not speaking.

Using a Noise Filtering on Sound Files: The Noise Reduction process filters a broad spectrum of frequencies; however, it is possible to remove specific frequency groups by applying a Band Pass filter.

Select the entire file by double-clicking on the wave form. Under TRANSFORM in the menu, select FILTERS and then FFT FILTER. In the FFT FILTER window, select LOW PASS 4000 Hz in the

PRESETS field. Click on the PREVIEW button, and then with the mouse pointer over the top yellow square of the graph that is above 4000Hz click and hold, and then slowly slide the square to the left. There may also be a need to slide the lower square to around 2000Hz.

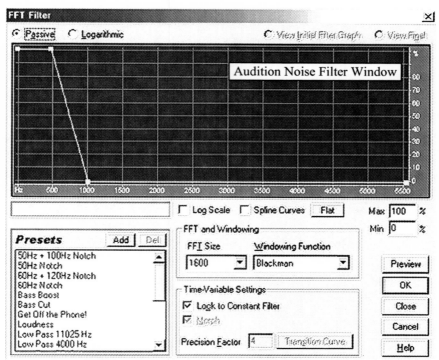

A corresponding change in the fidelity of the sound track can be heard as the filter settings are moved with the mouse and released. It is good to try a variety of settings and filters until an optimum setting is found. Remember that there is the UNDO capability if you go too far. If you place the pointer over the line of the graph and click, a new yellow square is formed. By doing this, the curve is shaped as needed.

Other than AMPLIFY, the procedures described above are probably the most commonly used in EVP analysis. There are a number of parameters in each procedure that should be experimented with to learn their full potential. Also, change the view between WAVEFORM and SPECTRAL to see how the frequencies are distributed in each part of the sound file. See how much power the frequencies have by using the FREQUENCY ANALYSIS function under menu item, ANALYZE.

There are other capabilities available in sound editing programs for analyzing sound files for EVP. For instance, words in EVP are sometimes spoken faster or slower than normal human speech. Under menu items TRANSFORM, TIME/PITCH, STRETCH, how fast the words are spoken can be changed. It is a good idea to load a sound file into the sound editor and simply experiment with each function. No harm can be done to a file as long as you do not save the file or if you only save the file using SAVE AS. Remember to have the UNDO feature always on.

A Word of Caution about "Enhancing" Sound Files

EVP are formed from background noise, rather than from a set of frequencies formed by the human vocal cords, throat and mouth. This means that many of the cues we depend on to recognize words are missing. This also means that word recognition in EVP is often dependent on the situation in which the words were recorded and the expectation of the listener. Thus, without the contextual cues, sometimes even a slight reduction of noise can change the way a sample is understood. So, be on the lookout for such changes and understand that some sound file enhancements may make it more or less difficult for others to understand EVP examples.

Management of Sound Files

The standard amongst AA-EVP members for sound files is the ".wav" format. The MP3 format is occasionally used, but usually only when the ".wav" file is too large to be attached to an email. As a courtesy to the other person, select only the portion of a sound file that has the EVP and save it as a wav file. Sound files can be huge and most email services will not accept very large attachments.

It is always good advice for people who work with EVP on their computer to have a way to backup the files outside of their computer. A read/write compact disc drive is excellent for this purpose. This tool will also make it possible for you to back up your personal information. The external read/write disc drives that connect to the computer via a USB port are fast enough for normal use.

American Association of Electronic Voice Phenomena (AA-EVP)

The American Association of Electronic Voice Phenomena (AA-EVP) was established in 1982 to provide objective evidence that we survive death in an individual conscious state. The main focus of the Association is on communication that comes from other dimensions through various technologies and ways of applying information gained through those communications to daily living. The Association is a Federally recognized 501 (c)(3) nonprofit organization. Despite its name, the Association is not just an "American" Association, as it includes members from around the world. The membership includes people who have experienced communication through many different types of devices, researchers who attempt to understand the "how" of these phenomena, researchers who are working on improving devices for communication, people who are seeking to communicate with loved ones and many others who do not experiment but wish to learn about these phenomena. The AA-EVP is not about religion.

The AA-EVP publishes a quarterly NewsJournal. The NewsJournal carries articles about what various researchers around the world are experiencing with these phenomena and the discoveries they are making in their research. It also carries articles that are aimed at helping people learn about the technology used to capture these phenomena. The Association has a website (aaevp.com), an optional email sharing group and an optional cross-country list, in which members can share their addresses with other members. There is also an online member archive that will eventually have all past Association NewsJournals as well as other important documents related to this subject.

Benefits of Membership in the AA-EVP

Learning to collect examples of these phenomena is not difficult, but there are many aspects of working in this field that can be difficult to work with alone. This book has told all that you need to know to begin experimenting and to understand what it is you will experience. However, we have found that there are always new questions and points of confusions that are expressed by people new to the field. Membership benefits of the AA-EVP are designed to help new people become veterans. For instance, the discussion board, we call the AA-EVP Idea Exchange, is proving to be a valuable means by which members can help one another. It is common for a member to post an EVP sample to the group to ask others what they hear. Other members often work with these samples using Audition, or another sound editing program, to enhance the sample, often making it more easily understood.

The AA-EVP online document Archive is another important benefit of membership. Eventually, all of the AA-EVP NewsJournals published since 1982 will be in the Archive as searchable text. Also, the *Spirit Voices* newsletters published by Bill Weisensale are being included. Bill's newsletters contain many important technical discussions about EVP. Access to this Archive provides an important history for the field, since many of the questions and ideas for experimentation that are discussed today have been discussed in the past. Often the results of those discussions are in the old NewsJournals. Historical documents will continue to be added to the archive as they become available to the Association.

Join the AA-EVP

Annual Membership dues for the AA-EVP are $30 in the United States. International members not receiving the NewsJournal via email must add $8.00. There is also a Sustaining Member membership of $100+ annually. Annual dues cover the cost of publishing the News-Journal. The Association relies on donations for all other operating costs. To join the Association copy, fill out and send the membership form at the end of this book, along with your annual dues, to AA-EVP, P.O. Box 13111, Reno NV, 89507. You will also find a membership form at aaevp.com. This can be printed and mailed to the above address or you can join online at aaevp.com via PayPal if you have a PayPal account.

References

1. AA-EVP Archive—A benefit of membership in the AA-EVP. The online Archive contains past AA-EVP NewsJournals and other documents of historical interest in the field of EVP and ITC.

2. Raudive, Konstantin—*Breakthrough: An Amazing Experiment in Electronic Communication with the Dead,* New York: Taplinger, 1971. First published by Colin Smythe, Ltd. and still available at www.colinsmythe.com.

3. Estep, Sarah—*Voices of Eternity,* Fawcett Gold Medal Book, Ballantine Books, New York, 1988.

4. Monroe Institute—62 Roberts Mountain Road, Faber, Virginia 22938

5. Light Institute—Galisteo New Mexico, www.lightinstitute.com.

6. Delphi University—P.O. Box 70, 940 Old Silvermine Road, McCaysville, GA 30555, www.delphi-center.com

7. *Psychic News*—British publication, Coach House, Stansted Hall, Stansted, Essex, CM24 8UD, www.snu.org.uk/p_main.htm.

8. Church of the Living Spirit—Chartered Auxiliary of the National Spiritualist Association of Churches,[9] The Reverends Sandra and Gene Pfortmiller, NST, Pastors, 3521 W. Topeka, Glendale, AZ 85308-2325, (623)-581-5544. Call about services.

9. The National Spiritualist Association of Churches—PO Box 217, Lily Dale, NY 14752-0217, (716) 595-2000, www.nsac.org. See also, www.nsacphenomena.net/

10. The Morris Pratt Institute—11811 Watertown Plank Road, Milwaukee, Wisconsin 53226-3342, (414) 774-2994, www.morrispratt.org.

11. Blank

12. "Messages from A Dead Soldier," *ARPR Bulletin,* June 2003, Depoe Bay, Oregon. Our thanks to The Academy of Religion and Psychic Research, and the Bulletin's editor, Michael E. Tymn.

13. Jürgenson, Friedrich—*Voice Transmissions with the Deceased,* (German to English Translation, T. Wingert & G. Wynne, 2001) Friedrich Jürgenson Foundation, Sweden (Original work published 1964) www.fargfabriken.se/fjf/

14. The Seth Books—information channeled by Jane Roberts. One noteworthy example is *Seth Speaks,* Bantam Books, 1974.

15. Doyle, Arthur Conan—*The History of Spiritualism,* Arno Press, New York, 1975.

16. Harold Sherman—*The Dead Are Alive,* Fawcett Gold Medal, New York, 1986.

17. Macy, Mark and Dr. Pat Kubis—*Conversations Beyond the Light,* Griffin Publishing, Irvine, CA, in conjunction with Continuing Life Research, Boulder, CO, 1995. You can learn more about Mark Macy at www.worlditc.org.

18. Audition—Sound editing software for PC by Adobe, www.adobe.com/products/audition/main.html. Was Cool Edit

19. Schrieber, Klaus— www.worlditc.org/a_12_itc_history.htm#Klaus%20Schreiber

20. Canon, Inc.—100 Jamesburg Road, Jamesburg, NJ 08831,

21. Pinnacle Systems—280 N. Bernardo Ave., Mountain View, CA 94043, www.pinnaclesys.com

22. Adobe Systems—San Jose Corporate Headquarters, Adobe Systems Incorporated, 345 Park Avenue, San Jose, California 95110-2704. www.adobe.com/

23. *Face Recognition in the Fusiform Gyrus*—Brown University research, www.brown.edu/Administration/News_Bureau/1998-99/98-154.html

24. Mitchell, Edgar, Sc.D. (for instance)—"Nature's Mind: the quantum hologram," Article at www.edmitchellapollo14.com/articles.htm

25. SORRAT—The Society for Research in Rapport and Telekinesis (SORRAT), Directed by Dr. Thomas Richards 309 West Ninth Street, Rolla, Missouri 65401

26. Wiseman, R. & Schlitz, M.—"Experimenter Effect and the Remote Detection of Staring," *Journal of Parapsychology*, 61, 197-208, (1998)

27. For Example—Stanley Sobottka, Emeritus Professor of Physics, University of Virginia, Charlottesville, VA 22904-4714, http://faculty.virginia.edu/consciousness/home.html

28. Clear Voice Denoiser—Speech Technology Center, software program designed to reduce noise in a sound track, www.speechpro.com/eng/products/denoiserkit.html.

29. Zammit, Victor—*A Lawyer Presents the Case for the Afterlife,* P.O. Box 1810 Dee Why 2099, NSW, Australia, www.victorzammit.com/book/

30. MacRae, Alexander—*The Mystery of the Voices,* Self published CD, 2000, Portree Skye, Scotland. See http://aspsite.tripod.com/ for details about the Alpha Device.

31. Presi, Paolo—Italian ITC researcher with Il Laboratorio, Bologna, Italy, www.laboratorio.too.it/

32. Butler, Lisa—Private recording, 2001

33. Talbot, Michael—*The Holographic Universe,* HarperCollins Publishers, New York, 1992

34. Rinaldi, Sonia—Brazilian ITC researcher, www.anttci.hpg.ig.com.br/ingles.html

35. Blanc-Garin, Jacque—Co-director of French EVP association, Infinitude. Conducted experiments with prearranged sleep and recording times. Results appear to affirm that some EVP may be thoughts of living people. www.chez.com/infinitude/Garde/EN_Garde.htm

36. Oberding, Janice—*Haunted Nevada,* Universal Publishers/uPUBLISH.com, Reno, Nevada, 2001. http://hauntednevada.com.

37. The Nevada Ghosts and Hauntings Research Society— www.ghrs.org/nevada/

38. American Ghost Society— www.prairieghosts.com

39. Spottiswoode, S. James P.—"Apparent Association between Effect Size in Free Response Anomalous Cognition Experiments and Local Sidereal Time," *Journal of Scientific Exploration,* Vol. II, No. 2, 1997, Lawrence, KS, www.jsasoc.com/library.html.

40. Konstantinos—*Contact the Other Side,* Llewellyn Publications, St. Paul, MN 55164-0383, 2001.

41. Anderson, George— well-known spiritual medium, www.georgeanderson.com

42. Botkin, Allan L., Psy.D—"The Induction of After-Death Communications Utilizing Eye-Movement Desensitization and Reprocessing: A New Discovery," *Journal of Near-Death Studies*, Vol. 18 #3, Spring 2000, IANDS, PO Box 502, East Windsor Hill, CT 06028, www.iands.org. Dr. Botkin's website is www.induced-adc.com/.

43. Smythe, Colin—Publisher, Colin Smythe Ltd, P.O.Box 6, Gerrards Cross, Buckinghamshire SL9 8XA, UK, http://www.colin-smythe.com/authors/covers/voices.htm

44. Stemmen, Roy—*Spirits and Spirit Worlds,* Aldus Books, London, 1975

45. Rogo, D Scott and Raymond Bayless—*Phone Calls From The Dead*, Prentice-Hall, Inc., New Jersey, 1979.

46. *Light Journal,* F.R. "Telephonic Communication with the Next World," August 20, 1921.

47. *Scientific American*—Scientific American, Inc. New York, NY, www.sciam.com/.

48. *The Journal of the American Society for Psychical Research*—5 West 73rd Street, Street, New York, NY 10023, www.aspr.com.

49. Ostrander, Sheila and Lynn Schroeder— *Handbook of Psychic Discoveries,* Berkeley Publishing Corp., New York, NY, 1975.

50. Bander, Peter—*Voices From the Tapes: Recordings from the Other World,* Drake Publishers Inc., New York, 1973. Initial German Language title: *Carry on Talking.*

51. Fuller, John G.—*Ghost of 29 Megacycles, A New Breakthrough in Life after Death?* Souvenir Press, London GB, 1985.

52. Webster, Kenneth—*The Vertical Plane,* Rare Publishers Ltd., London, 1989.

53. *VTF Post*—German EVP and ITC organization, Vereins Fur Tonbandstimmenforschung (VTF), www.vtf.de

54. *AA-EVP NewsJournal*—American Association of Electronic Voice Phenomena, Reno, NV, 1982-2003. www.aaevp.com. Please note that the name of the AA-EVP newsletter was changed to the *AA-EVP NewsJournal* in 2003.

55. Stevenson, Dr. Ian—Children's Past Life Research Center, www.childpastlives.org.

56. *Transdimension*—International Network for Instrumental Transcommunication (INIT), Published in the United States by Continuing Life Research, Mark Macy, Boulder CO, January-June 1999.

57. Chisholm, Judith—*Voices From Paradise: how the dead speak to us,* Jon Carpenter Publishing, 2000, Kent, UK.

58. Ebon, Martin—*Communicating with the Dead,* The New American Library, New York, 1968.

59. Cabral, Euvaldo—Brazilian EVP and ITC researcher in the field of extremely low signal to noise ratio speech enhancement techniques who is now associated with the Noetics Institute, Incorporated, Plainfield IN.

60. Robb, Stewart—*Strange Prophecies that Came True,* Ace Books, New York, 1967.

61. *Journal of the Society for Psychical Research*—49 Marloes Road, Londan, W8 6LA, www.spr.ac.uk, Tel.: 020-7937-8984.

62. *Info News*—English edition of *Cercle D'etudes sur la Transcommunicaton, Luxembourg,* (Luxembourg Study Circle in Transcommunication,) Translation: Hans Heckman, US Publishing: Continuing Life Research, PO Box 11036, Boulder, CO, 80301. (No longer published.)

63. Feola, Jose, Ph.D—"The Alpha Mystery," *Fate* Magazine, July 2000, PO Box 64383, St. Paul, MN 55164-0383.

64. Palmer, Dale—Noetics Institute, Inc., Plainfield IN.

65. *Contact*—Published by Continuing Life Research, PO Box 11036, Boulder, CO, 80301. (No longer published.)

66. Solomon, Grant & Jane—*The Scole Experiment,* Piatkus (Publishers) Ltd, 1999, London.

67. "Scole Report"–*Proceedings*, Society for Psychical Research, Vol. 58, Pt 220, pages 150-392 with supplementary criticisms and responses by Montague Keen. http://moebius.psy.ed.ac.uk/~spr/scole_report.html

68. *ITC Journal*—Published by Dr. Anabela Cardoso, Apartado de Correos 3157 VIGO - Pontevedra – Spain.

69. Charrous, Robert—*The Mysterious Past,* Berkley Medallion, NY, 1975, Page 253.

70. *Fate Magazine*—PO Box 460, Lakeville MN 55044-0460

71. Schwartz, Gary, Ph.D. and William L. Simon—*The Afterlife Experiments, Breakthrough Scientific Evidence of Life After Death*, Pocket Star, New York, 2002

72. Bion, Stephen—EVPMaker, Personal Computer program developed by Bion as a software-based EVP experimental device. See www.stefanbion.de/evpmaker/.

73. *Spirit Voices*—Technical EVP newsletter published by Bill Weisensale, Sandy Valley, NV, 1980 to 1995. (Issues are being included in the AA-EVP Archive.)

74. *Psychic Observer and Chimes*—Journal of Spiritual Science, Vol. XXXVII No. 5, ESPress, Inc., Washington DC, 1977.

75. Festa, Mario Salvatore—"A particular experiment at the psychophonic centre in Grosseto, directed by Marcello Bacci," *ITC Journal*[68] No.10 June 2002

76. *The Spiritual Scientist*—Spiritual Science Foundation, Scole, Diss Norfolk, www.psisci.f9.co.up.

77. *Le Messager*—tri-annual magazine published by the French ITC Association, Infinitude, Les Mesnil des Frétils F-27250 Les Bottereaux. We thank Jacques Blanc-Garin of Infinitude for translations.

78. Delaware Valley Demonology Research—
    www.demonologyresearch.com/DVDR/evps.htm

79. The Silva Method—A self-improvement training program,
    www.silvamethod.com.

80. Builders of the Adytom (BOTA)—An Ancient Wisdom School based on
    the Hermetic teachings, www.bota.org/.

81. Rosicrucians Order AMORC—An Ancient Wisdom School based on the
    Hermetic teachings, www.rosicrucian.org/.

82. Botkin, Dr. Allan L.—*Reconnections: A Psychological Discovery for
    Resolving Grief and Traumatic Loss,* (Book is not yet available at the
    time of this writing), http://induced-adc.com/iadctherapy.htm

83. Reiki—A system of energy or spiritual healing, www.reiki.org/.

84. Flint, Leslie—*Voices in the Dark,* Two Worlds Publishing Co Ltd., London 2000.

Index

About the Title

In 1848, the Fox sisters experienced persistent raps and other physical phenomena in their family cottage in Hydesville, New York. One evening, one of the Fox sisters attempted to mimic the rapping sound by snapping her fingers. The youngest sister, Catherine, said, "Mr. Split-foot, do as I do," and clapped her hands a number of times. The unseen entity rapped the same number of times. The entity turned out to be a peddler who had been murdered in the cottage years before. His name was Charles B. Rosna, and he was important in that he persisted in establishing communication with the Fox family. In doing so, he gave proof that life continues beyond physical death. It is this exchange between the Fox Family and Mr. Rosna that is credited as the beginning of Modern Spiritualism.

Using the alphabet as a key for interpreting the raps, the first message from the peddler was, "Dear Friends, you must proclaim these truths to the world. This is the dawning of a new era: you must not try to conceal it any longer. When you do your duty, God will protect you and the good spirits will watch over you." Thus, the Hydesville raps proclaimed to the world that, "There is no death and there are no dead."

From information Spiritualist historian The Reverend Marilyn Awtry-Smith has provided, we believe that the phrase, "There is no death and there are no dead," was coined by Mercy E. Cadwallader when she had the tombstone engraved and placed in the yard of the Fox Cottage. Cadwallader was given a four-line song from spirit that ended in, "There is no Death, There are no Dead, We live, and love you still."

As an early figure in Spiritualism, Mercy E. Cadwallader is known as the Animated Encyclopedia of Spiritualism. She is author of the booklet, *Hydesville in History.*

Footnote of History: It is a matter of record that the older Fox Sister, Leah, became at odds with the two younger sisters, Margaretta and Kate. The root of this problem was Kate's drinking and a subsequent threat supported by Leah to take Kate's children from her. The circumstances of an established church's desire to discredit Spiritualism, and a reporter eager to facilitate this cause by capitalizing on the rift between the Fox Sisters, ultimately resulted in Kate's public admission that she had faked the famous raps by cracking a double-jointed big toe. She later recanted her confession in a well documented, written statement.[15]

The historical record of the interaction between the peddler and the Fox Sisters has been collaborated by physical evidence and careful evaluation by trained researchers. This research has set aside an obvious effort to debunk an important historical event.

AA-EVP Membership Form

Membership in the Association is open to anyone. Annual dues are:

0 Member	$30	All services for one year
0 * International Member not using Email	$40	All services for one year
0 Sustaining member	$100	Member + name listed in NewsJournal

* International member **not receiving NewsJournal via email as a PDF file.**

All dues must be in US Denomination. Please make checks payable to AA-EVP.

Mail check and form to: AA-EVP, PO Box 13111, Reno, NV 89514

Name_____

Email Address (Optional)_____

Phone Number (Optional)_____

Address_____

City/State/Zip code_____ Country_____

Do you wish to be on the Cross-country list?_____

   Do you wish to include your physical address?_____

   Do you wish to include your email address?_____

Do you wish to receive the NewsJournal via email _____
or, via the Postal Service?_____

On the other side, tell us a little about yourself, and if you record and what techniques you like to use.

Member Profile (Please check what best describes your interest in EVP)
[ ] I record on a regular basis.
[ ] I plan to begin recording.
[ ] I joined because of the loss of a loved one.
[ ] None of the above, I have an interest in EVP and its evidence for survival.

I understand that the cross-country list should not be used for commercial purposes or the furtherance of personal causes. By indicating that I want to share my name and address with others through the cross-country list, I realize this is a private list and I agree that other names on the list will not be given to anyone who is not on the list. I also understand that my name will be removed from the list and my membership in AA-EVP will be terminated if I violate this agreement.

Signed _____Date_____

You can now submit a membership form online at aaevp.com. You can also submit your membership dues online via PayPal. You will be guided to do so after filling out the online membership form. The AA-EVP is a 501 (c)(3) organization. Funds in excess of dues are tax deductible in the USA.

Ordering Additional Copies of This Book

Copies of this book may be ordered from the AA-EVP at http://book.aaevp.com. Credit card orders are accepted on the website via secure PayPal.com. You can order from AA-EVP, PO Box 13111, Reno, NV 89507 using a check or money order made payable to AA-EVP. Allow additional six to ten days for personal checks to clear.

There is No Death and There are No Dead
$18.00 each

Number of copies: _____ X $18.00

Subtotal: _____

Nevada State Sales Tax: _____ (Nevada Residence only)

Shipping and Handling: _____

Total Enclosed: _____ Please make check to: AA-EVP

Shipping and Handling Charges

	First Class	International
First book:	$6.00	$10.00
Each additional:	$2.00	$10.00

Please attach a list with instructions if you would like signed copies of this book.

Mail book(s) to: Name: _____

Street address: _____

City, State, Zip: _____

Telephone Number: _____

Email Address (Optional): _____

Contact aaevpsupport@aol.com if you have questions.

Printed in the United States
218036BV00007B/8/A